T0173159

THE NATURAL HISTORY MUSEUM BOOK OF

ROCKS & MINERALS

A CONCISE REFERENCE GUIDE

CHRIS & HELEN PELLANT

Published by the Natural History Museum, London

First published by the Natural History Museum,
Cromwell Road, London SW7 5BD

ISBN 978 0 565 09505 5

A catalogue record for this book is available from the British Library.

10 9 8 7 6 5 4 3 2

Design by Mercer Design, London
Reproduction by Saxon Digital Services, Norfolk
Printing by Toppan Leefung Printing Limited, China

Contents

Introduction

Rocks and minerals are the basic materials which form the Earth's crust. The properties and occurrence of both are dealt with in this book. Each mineral and rock is accompanied by a photograph of a specimen and a written description, both of which are an aid to identification.

The Earth's landscapes are very much related to rock types and the way weathering and erosion affect them. Some rocks are more resistant than others to these processes. The metamorphic rock, gneiss, for example, and the igneous rock, dolerite, are relatively tough and usually not greatly affected. On the other hand, limestones, with their bedding planes and joint systems and susceptibility to chemical weathering, can give rise to distinctive landscapes, with cave systems, underground water courses and limestone pavements.

Rocks have provided building materials for thousands of years. Slate, because of its cleavage into thin plates, is used for roofing, clay for brick and tile manufacture, limestone for cement making and attractive rocks such as granite, marble and larvikite are employed as ornamental facing stones on the fronts of buildings.

Sandstone, a very porous rock, often serves as an aquifer, providing an underground water supply. It can also be a trap for oil and gas. Rock salt provides material for the chemical and food production industries. Coal, which powered the Industrial Revolution, is still widely used around the world, but this practice is now being phased out, as the burning of fossil fuels is a major source of global warming.

The metals which are so important in countless aspects of daily life are derived from minerals. Valuable ore minerals include magnetite and

hematite (iron), sphalerite (zinc), galena (lead), chalcopyrite (copper) and cassiterite (tin). The nuclear industry relies on uranium, found in the mineral uraninite. Small modern batteries often contain the element lithium, mainly obtained from spodumene. Gypsum is used in the manufacture of plasterboard, fluorite as a flux in the production of steel and polyhalite is an important fertilizer.

Many minerals have fine crystal form and striking colours, making them well worth collecting. There are a number of dealers who sell rock and mineral specimens, but it is quite possible for an individual to collect them in the field. Mine and quarry spoil heaps are good places to look and exposures such as cliffs, at the coast and inland, old quarries and cuttings may provide specimens. By using guidebooks, maps, and the internet, it should be possible to pinpoint locations to visit. Permission should always be sought before going onto private land. Personal safety is important. Cliff and quarry faces can be dangerous, and even a hard hat offers little protection from falling rocks. A geological hammer should only be used for breaking up fallen material. Goggles will protect the eyes from rock splinters. Restricting the number of specimens collected from any one site is essential. Specimens need to be wrapped before taking them away. It is important to make a detailed record of the location where a specimen was found; material without a location has little scientific value. Before adding specimens to a collection, they should be carefully cleaned, labelled and catalogued.

◀ Chrysocolla
from Santa Fe Mine,
Chiapas, Mexico.

Introduction to rocks

A rock is defined as an aggregate of minerals or mineral particles. There are commonly very few minerals making up any rock. Hard crystalline rocks, loose sand and soft, sticky clay are all rocks. The mineralogy and other features of a rock are a direct result of how and where it formed. Petrologists (geologists who study rocks) seek to understand not only how rocks have been created, but also the sequence in which they formed.

There are three broad categories into which rocks are classified based on their origins and features. These groups help us to understand and identify rocks, but there can be some overlap. Igneous, metamorphic and sedimentary rocks are related by a cycle of formation. Igneous rocks are of primary origin and form as molten magma or lava solidifies in the Earth's crust or on the surface. When these rocks are weathered and eroded, the resulting grains are deposited to form sedimentary rocks. Both igneous and sedimentary rocks can be altered by direct heating (contact with magma), or by the pressures and other effects of plate tectonic movement, to form metamorphic rocks. Burial and subsequent heating of any of these rocks may cause melting and the cycle starts again.

IGNEOUS ROCKS

Composition

The composition of an igneous rock depends on where the rock formed and the type of magma from which it cooled. Magma from deep in the crust, or in the upper parts of the Earth's mantle, forms rocks with a relatively low silica content. Such rocks tend to occur in oceanic regions, where the crust is thinner, and are known as 'basic' rocks. Magma found in the thick continental crust generally has a higher proportion of silica

◀ Pegmatite from Kodarma, India.

and rocks from this region are known as 'acid' rocks. Though the terms 'acid' and 'basic' are taken directly from chemistry, in the geological context they refer to the total silica content of the rock. Acid rocks have more than 65% total silica and usually over 10% quartz; basic rocks contain between 45% and 55% total silica, with less than 10% quartz. Intermediate rocks are between these two and the ultrabasic rocks contain less than 45% total silica. There are relatively few important rock-forming minerals in the igneous rocks and these are mainly silicates. Acid rocks are rich in feldspars, micas, hornblende and quartz, while basic rocks are mainly composed of feldspars, pyroxenes and olivine. Acid rocks tend to be pale coloured and have a specific gravity of around 2.6, whereas basic rocks are a darker colour, with a higher specific gravity of about 3.2. Intermediate rocks have features between the acid and basic rocks, while the ultrabasic rocks are mainly composed of dark ferro-magnesian minerals such as pyroxenes and olivine, which give them a high specific gravity of around 3.5.

Grain size, texture and structure

The grain size of an igneous rock is determined by its cooling history. If magma cools slowly at depth, in a large mass such as a pluton or batholith, there will be considerable time for mineral crystals to develop. Some batholiths may take tens of millions of years to consolidate. Crystals (grains) in these rocks can be over 5 mm (⅕ in) in size and the rock is then said to be coarse-grained.

If the magma creates structures such as dykes and sills, then cooling is faster, and smaller crystals are formed, usually between 5 mm (⅕ in) and 0.5 mm (¹⁄₅₀ in) size. These rocks are medium-grained. Lava tends to solidify very quickly, as it is extruded onto the relatively cold surface of the Earth, and at times under water, and crystals form that are smaller still, often not large enough to be seen with a x10 hand lens. These grains, less than 0.5 mm (¹⁄₅₀ in) size, make up the fine-grained rocks.

Rock texture is the relationship between the size, shape and position of the grains in a rock. For example, when magma cools slowly, euhedral (well-formed) crystals can develop, but rapid cooling can produce anhedral (poorly shaped) crystals. Equigranular rocks have similar-sized crystals, but rocks with a porphyritic texture have relatively large crystals set in a finer groundmass. This texture is produced when the temperature is maintained for some time and relatively large crystals develop in the molten rock. The remaining magma may then cool rapidly around these crystals, producing the porphyritic texture. Many other textures exist. A poikilitic texture has relatively large crystals, with smaller ones of different minerals within them, and an ophitic texture, not uncommon in basic rocks such as gabbro, can develop when plagioclase crystals are enclosed in larger masses of pyroxene.

Structural features in igneous rocks may only be seen in field exposures, though some are visible in hand specimens. Lava often contains the evidence of gas bubbles in the original molten rock. This evidence may be in the form of small, rounded cavities called vesicles. These are sites where mineralization can readily occur. When they are infilled, the term amygdale is applied and the lava is then called amygdaloidal lava. Lava and some intrusive rocks can have flow banding (see p.16 rhyolite). Masses of included rock, which are different from the composition of the host rock, are referred to as xenoliths. These are usually more common around the margins of an intrusion and represent fragments of the local 'country rock' caught up in the magma. A large-scale structure, common in some lavas and small intrusions, is columnar jointing. This develops at right angles to cooling surfaces. Dykes are small-scale discordant intrusions, which cut across local rock structures. Columnar jointing in them is often almost horizontal. Sills follow existing features, such as sedimentary bedding, and any columnar jointing, as in lava flows, tends to be vertical.

▼ Ropy lava flow exposed in the Thingvellir National Park, Iceland. This area sits on the mid-Atlantic rift where the North American and Eurasian plates move apart. Earthquakes and lava eruptions occur as the plates move.

METAMORPHIC ROCKS

These rocks have been altered from their original state by various processes in and on the Earth's crust. Metamorphic rocks can have totally different features from the rocks from which they formed. Metamorphism is caused by high temperature, elevated pressure, the effects of hot fluids or combinations of these. No melting takes place and alterations occur in the solid rock. For this reason minerals are often different from those in igneous rocks, where they form within liquid magma. Three main groups of metamorphic rocks are identified: regional, contact and dynamic.

▲ Gneiss exposed at Achmelvich on the north-west coast of Sutherland, Scotland. The alternating dark- and pale-coloured banding, typical of gneiss, is obvious.

Regional metamorphism

Plate tectonics theory suggests that the movement of parts of the Earth's crust creates mountain ranges. Preformed rocks, which are subjected to the pressures, temperatures and folding processes of mountain formation, are affected by regional metamorphism. As the name suggests, this type of metamorphism can occur over wide areas. Different grades of regional metamorphism are recognized, though often there is no clear demarcation between one grade and another. Temperature is only an important factor at higher grades of metamorphism, commonly reaching between 200°C (328°F) and 700°C (1,228°F). Pressure increases with folding and depth; for example, at about 20 km (12 miles) depth, the pressure is over 6,000 times the surface pressure.

Low-grade regional metamorphism occurs at no great depth in the crust, where pressures are low and temperatures are of little consequence. Only rocks such as sedimentary shale and clay, along with volcanic ash, are readily altered at this grade.

These rocks are changed to the metamorphic rock slate. The main feature of this rock is its slaty cleavage. This is a structural feature related to the pressure associated with metamorphism affecting the small-sized mineral particles, especially flaky minerals such as mica, in the original rock. These particles become aligned, giving slate a characteristic cleavage, and the rock easily breaks along its cleavage planes. 'Slaty cleavage', as it is known, generally forms at right angles to the direction of greatest pressure. Features of the pre-metamorphosed rocks, such as bedding planes and fossils (which may be stretched and distorted), can often be seen.

Medium-grade metamorphism affects a wider variety of rocks than low-grade metamorphism, generally because conditions are more extreme. Here, temperature is an important influence and pressure is greater. Slate itself can be altered by medium-grade metamorphism, as can certain igneous rocks and many sedimentary rocks, including sandstone and limestone. Medium-grade rocks differ structurally from those formed at low grade. Instead of the reasonably flat planes of slaty cleavage, a more wavy foliation, called schistosity, is produced. Schist is the typical rock of medium-grade metamorphism. This has grains of medium size, aligned along the planes of foliation, and it does not break as readily as slate.

New minerals may develop at this grade – kyanite, garnet and pyroxenes being typical – and mica is often found on the foliation planes, giving the rock a characteristic sheen. Certain original features, such as bedding, may still be visible.

High-grade metamorphic rocks are formed when both temperature and pressure are extreme, deep within the roots of mountain belts. Just about any rock, including granite and other igneous rocks, can be metamorphosed in these regions. High-temperature fluids, which move at depth in the crust, can change the chemical components of rocks, which may become plastic under high-grade influences. The typical rock is gneiss. This has a characteristic structure of alternating light and dark bands, where different minerals have become segregated. The paler bands are rich in quartz and feldspar, while in the darker bands, biotite mica, pyroxenes and amphiboles are common. Augen gneisses have rounded segregations of minerals set in their overall banding. As the conditions of high-grade metamorphism develop, melting of the rocks involved may occur, so that magma is created. Gneiss and related high-grade rocks are commonly of great age. Many regions of the pre-Cambrian continental crust, such as the Canadian Shield, are composed of these rocks.

Contact or thermal metamorphism

Extreme temperature is the main influence in this type of metamorphism. Magma or lava may be at temperatures of over 1,000°C (1,768 °F) and this molten igneous rock changes any preformed rocks with which it is in contact. This change is usually mineralogical but can also be structural. Recrystallization takes place, often with the addition of new minerals. Sedimentary rocks such as sandstone and limestone are altered by recrystallization into metaquartzite and marble. Slate may lose its cleavage and become tough, flinty hornfels without structure. Even bedding and other sedimentary features are removed by recrystallization. Impurities in the original rocks may be changed to new minerals; marble, for example, often contains brucite and diopside.

The area around a body of igneous rock that is altered by contact metamorphism is called the metamorphic aureole. This varies in extent according to both the size of the intrusion and the type of magma. At its outward margins, the metamorphosed rock is gradually replaced by unaltered country rock (the rock in that area). Where a large mass of igneous rock, such as a granitic batholith, has intruded country rock, the temperature near to the magma can be as high as 700°C (1,228°F), but around 4 km (2½ miles) from the magma body, it may fall to about 350°C (598°F). Granitic magma may take thousands or, in extreme cases, millions of years to consolidate. This type of magma is associated with high-temperature fluids, which can affect the country rock by adding and removing chemicals. Smaller intrusions, such as dykes and sills, tend to have a narrower metamorphic aureole. Lava flows produce contact metamorphism in the rocks over which they spread. In the field, even though they may be composed of very similar rock, a lava flow can be distinguished from a sill, as its metamorphic aureole is only present on one side, whereas a sill has aureoles on both sides.

Dynamic metamorphism

This type of metamorphism is localized and not as common as contact and regional metamorphism. It is associated with large-scale fault movements, where thrusting has caused great masses of rock to move along a low-angled fault plane. Rock masses may be moved many tens of kilometres along such planes, causing older rock from depth to lie above younger rocks. Where rocks are broken and pulverized near the fault plane, dynamic metamorphism occurs, producing mylonite, a rock with streaked-out features. This has a fine groundmass and can have a porphyroblastic texture, containing larger, more resistant fragments.

SEDIMENTARY ROCKS

Sedimentary rocks are usually formed on the Earth's surface and so their creation can be readily studied and understood. Many result from the processes of erosion and weathering, which produce fragmentation of sediment, transportation of this material by water, ice and the wind, and finally deposition in water or on dry land. Some are formed of biological material, while others have a chemical origin. Sedimentary rocks are distinguished from igneous and metamorphic rocks by their characteristic structures such as stratification (bedding). Bedding planes divide one stratum (layer) from another. In addition, there is a great range of other sedimentary structures, including cross-bedding, graded bedding, ripple marks and desiccation cracks. Sedimentary rocks also often contain fossils, which provide information about evolution and are stratigraphically useful in determining relative age and correlating strata.

The shape and size of the grains that make up a sedimentary rock can be used in classification. Grains often become rounded during transportation and others may, in different environmental situations, have an angular shape. The main difference between coarse-grained conglomerate and breccia, for example, is that conglomerate, usually water deposited, has rounded fragments, whereas the fragments in breccia, which can be a scree deposit formed on land, are angular. The grains in a sedimentary rock are sorted by erosion and transportation. As they are carried, their size is reduced. A sedimentary rock that contains grains of much the same size is said to be well-sorted, while one with a variety of grain sizes is poorly sorted.

Loose sediment may be turned into a more consolidated rock in a variety of ways by the process of diagenesis. This can take place at no great depth in the crust and high temperatures and pressures are not involved. Chemical-rich fluids, including trapped seawater, can seep into sediment and deposit cementing minerals such as calcite and quartz, bonding the grains together. The weight of layers of sediment causes solution on the edges of grains and compaction, both processes removing pore spaces and aiding consolidation. As well as making the grains fit together more tightly, the weight of overburden helps to remove interstitial fluids (fluids that sit between the grains). There can be an 80% reduction in the thickness of sediment by burial, as pore spaces and fluids are reduced. This is most noticeable in fine-grained sediment, such as when mud turns to shale.

▼ The Green Bridge of Wales in Pembrokeshire, is a limestone coastal arch, eroded by wave action, with sedimentary strata dipping inland.

Detrital sedimentary rocks

These rocks are essentially formed by processes such as weathering, erosion, transportation and deposition. Weathering and erosion are often confused, though they are very different. Weathering is the breakdown of rock, in situ, without movement. Chemical weathering involves rainwater, which is slightly acidic. The acidity arises because rain naturally absorbs carbon dioxide from the atmosphere, forming carbonic acid. As the acidic rain falls on the rocks it causes a chemical reaction, especially in rocks that are rich in calcium carbonate. For example, it reacts with the insoluble calcite (calcium carbonate) in limestone, being altered to calcium bicarbonate, which is soluble in water and so is washed away. Other minerals may also be changed in similar ways. Feldspar in granite and other rocks, for example, is altered to clay. Weathering may also be mechanical, involving changes in temperature. Minerals in rocks expand and contract differently and this may cause disintegration, especially if the rocks have already been affected by other forms of weathering. In high latitudes and altitudes, the effects of repeated freezing and thawing and the pressure created when ice forms in rock joints can produce rock fragments which subsequently fall onto slopes as scree. Biological weathering, caused by plants and animals, includes expansion of rock joints by tree roots and the burrowing activities of many creatures.

Erosion is the decomposition of rock material during movement. This occurs in many surface environments, including glaciers, rivers and the sea. The wind can carry and erode rock fragments, often causing sand-sized particles to become rounded, unlike larger land-deposited fragments. Other processes may involve gravity, which causes movement downslope and fragmentation.

The grain size of a sedimentary rock tends to lessen the farther the rock is from the source area. Coarse-grained rocks, such as conglomerates and breccias, are often formed near their source area, but fine-grained sands and muds are frequently deposited from material carried out to sea. Farther from the source area, sediment may become more mature. Maturity is a quality related to the abundance of quartz in a sedimentary rock. During weathering, erosion and transportation, some of the minerals in the loose sediment are gradually removed, while the relative percentage of quartz, a hard and chemically resistant mineral, increases. Sediment that contains a high percentage of quartz is said to be mature. Feldspar content is another way to determine maturity. This mineral is readily weathered; a rock such as the sandstone arkose, which contains more than 25% feldspar, has been rapidly deposited and is said to be immature.

Organic sedimentary rocks

These sedimentary rocks, often limestones, contain a large amount of biological material, commonly derived from mollusc and brachiopod shells and corals. This material is frequently broken. The rocks may consist of a lime-rich mud containing larger fossils. Limestones are common in this category and can be named according to their main constituent, for example coral limestone or shelly limestone. Coal is another example of an organic sedimentary rock, being formed from plant material.

Chemical sedimentary rocks

These rocks are deposited by inorganic chemical processes, often by precipitation of minerals from aqueous solutions. In this category are sediments such as rock salt, rock gypsum, potash, marl and dolostone, deposited from evaporating saline waters. Certain limestones are also formed chemically. The small, rounded ooliths in oolitic limestone are composed of inorganically precipitated layers of calcite.

Igneous rocks

Igneous rocks are formed by the cooling and crystallisation of molten rock generated deep in the Earth's crust. When below the surface, this is called magma, and as it cools, intrusions such as batholiths, sills and dykes develop. Lava is magma which has been erupted onto the surface, producing extrusive landforms such as volcanoes.

Granite

Composition: Granite is classified as an acid igneous rock. This classification is based on its composition: overall the rock contains more than 65% total silica and over 10% quartz. The quartz content is frequently in excess of 20%, even as much as 30%, of the rock's composition. The bulk of granite is composed of the silicate minerals feldspar, mica and quartz. Quartz appears as small, greyish crystals, set among better formed feldspar and mica. The content of orthoclase feldspar exceeds plagioclase and both dark biotite and pale muscovite mica can also occur. The silicate amphibole mineral hornblende is frequently present and accessory minerals include beryl, tourmaline, fluorite and pyrite. The overall mineral content gives granite a paler colour than many other igneous rocks. Depending on the colour of the predominant feldspar in the rock, granite can be pink (with pink or reddish-coloured orthoclase) or white. If there is much dark mica and/or tourmaline, the rock may be darker and flecked with black.

▲ Granite containing dark biotite mica form Jersey, Channel Islands.

Grain size, texture and structure: Granite is composed of coarse-grained crystals over 5 mm ($^1/_5$ in) in diameter, which can be readily seen without a hand lens. Many can be much larger than this. Granites may vary in their texture. Some are equigranular, with all the constituent crystals being about the same size. Others have a porphyritic texture. This is produced when large crystals (phenocrysts), often of feldspar, are set into a rock groundmass of smaller crystals, as the different minerals in granite crystallize in sequence during cooling. The temperature at which feldspars form may exist for some time, allowing the crystals to become large, and then, as the remainder of the magma cools more quickly, these large crystals are trapped in the smaller-grained rock groundmass. If there is an intergrowth of feldspar and quartz, graphic granite is produced. Here, large feldspars contain within them slender quartz crystals, giving the impression of writing. Granites may be named pink, white, porphyritic and graphic granite because of predominant colouring or texture. Xenoliths of country rock are common in granite intrusions, especially near the margins.

Occurrence: Granite is a common igneous rock that occurs in large igneous intrusions (plutons and batholiths), often many kilometres in diameter. Originally, the magma from which granite crystallizes was intruded from depth into the upper parts of the Earth's crust, the highest parts of the intrusion being a few kilometres below the surface. Granite is only exposed on the surface today because of erosion and weathering of the original overlying rocks. Plutonic intrusions tend to occur in the continental crust, often in the roots of mountain chains. Many granite bodies are discordant and have relatively sharp contact with the surrounding rocks (country rocks). Heat from the magma often produces a metamorphic aureole. Some granite masses, however, show a gradual change from igneous to high-grade metamorphic rock. The relatively coarse grain size develops because of the slow cooling of the magma, which may take many millions of years to cool completely.

◀ Graphic granite from Central Provinces, India.

▶ Porphyritic granite form Shap, Cumbria, England, UK, containing large phenocrysts of orthoclase feldspar.

Pegmatite

Composition: Pegmatite is a silica-rich igneous rock of acid composition, similar to granite. It contains over 10% quartz and much feldspar, commonly orthoclase, though other feldspars can occur in the form of sodium-rich plagioclases. As in granite, mica is also a common mineral. The total silica content is over 65%, giving pegmatite an overall pale colour. The mineralogy can be varied and other constituents include hornblende, tourmaline, topaz, beryl, apatite, spodumene, titanite, rutile and zircon. Individual types of pegmatite may be named according to certain important minerals, for example, mica pegmatite, tourmaline pegmatite and feldspar pegmatite. The rock name pegmatite strictly refers to an acid igneous rock, but the adjective 'pegmatitic' is sometimes used to describe other igneous rocks of very coarse grain size.

Grain size, texture and structure: Pegmatite has very large crystals, which are easily seen with the naked eye. This rock usually has an equigranular texture, but phenocrysts can occur. Crystals over 10 m (33 ft) long have been found in some pegmatites.

Occurrence: Pegmatite forms in veins and small dyke intrusions, often around the edges of larger masses of granite. In this situation, remaining fluids in the late stages of magmatic crystallization cool slowly, allowing time for the large crystals in pegmatite to form. These fluids also contain the chemical materials from which rare minerals develop. Pegmatites are known for containing fine, well-formed mineral crystals.

▼ Pegmatite from Kodarma, India.

Micrograzite

Composition: An acid igneous rock with a total silica content of more than 65% and more than 10% quartz. Alkali feldspars (potassium and sodium feldspars) are important minerals, orthoclase being dominant over plagioclase. Both light-coloured muscovite mica and dark biotite are common. This mineralogy gives microgranite a pale overall colour; it can be pinkish if it has a high content of pink orthoclase.

◀ Microgranite from Cornwall, England, UK.

Grain size, texture and structure: Microgranite is a medium-grained igneous rock, with constituent crystals generally between 5 mm (⅕ in) and 0.5 mm (⅟₅₀ in) diameter. A hand lens is very useful when studying the mineralogy of the rock. The texture is commonly equigranular, with a mass of interlocking, euhedral crystals, but porphyritic microgranites (quartz porphyry) are well known. Often the phenocrysts in such rocks are of pale-coloured feldspar.

Occurrence: Microgranite forms in small intrusions such as dykes and sills and also around the more rapidly cooled margins of larger igneous bodies.

Rhyolite

Composition: This is an acid lava with similar overall composition to granite. It therefore contains over 65% total silica and more than 10% quartz. Orthoclase feldspar is more common in this rock than plagioclase, and mica is also present. Glass often occurs due to the lava's rapid cooling. Rhyolite is pale coloured, though it can vary from grey, green and brown to almost white.

▶ Rhyolite from County Antrim, Northern Ireland, UK.

Grain size, texture and structure: Rhyolite is very fine-grained (grains less than 0.5 mm or ⅟₅₀ in), so a hand lens is essential for observation of the mineral content. At times the rock has cooled so quickly that the minerals are anhedral (poorly shaped). Even with a microscope, identification can be difficult, especially when much glass is present. Being so fine-grained, rhyolite often has a flinty appearance. Viscous rhyolitic lava frequently produces a flow-banding structure and this curved and swirling pattern may be accentuated by bands of different colours and grain sizes. Rhyolite is commonly porphyritic, with small feldspar phenocrysts set in the very fine groundmass of the rock. These crystals may follow and emphasize the flow-banding structure. Rhyolite can contain vesicles (cavities): indeed, pumice is a highly vesicular type of rhyolite. A spherulitic texture sometimes occurs in rhyolites, with rounded structures composed of radiating needles of feldspar or quartz, formed in rapidly cooling lava.

Occurrence: Rhyolite is a rapidly cooled, highly viscous lava. Such lava eruptions are usually very explosive, the extruded lava only flowing short distances. Often the lava solidifies as a plug in the volcano's crater. Rhyolite may also occur in small dykes.

Obsidian (and snowflake obsidian)

Composition: An acid igneous rock, obsidian contains a high proportion of silica (well over 65% in total) and there is a corresponding high percentage of quartz. The mineralogy is difficult to observe, as the rock is made almost entirely of volcanic glass, which produces a vitreous lustre. Obsidian has the same chemical composition as rocks such as rhyolite and granite. Unlike these rocks, however, obsidian is very dark coloured, generally being black or dark green. In some specimens there are small phenocrysts, usually composed of feldspar or quartz. Snowflake obsidian is recognized by the many small, white patches that occur through the rock, contrasting with obsidian's black colour and vitreous lustre, giving the rock an attractive appearance. These 'snowflakes' are devitrified glass and masses of cristobalite, a mineral related to quartz.

Grain size, texture and structure: Even under the microscope, any crystals in obsidian can be very difficult to detect, since it is a glassy rock. The crystals that can be observed are generally anhedral (poorly formed). The texture is usually equigranular. Flow banding is occasionally visible. When broken, obsidian produces a very sharp-edged fracture, with curved (conchoidal) patterns on the broken surfaces. In the past, the rock has been used for making tools, because of its sharp fracture and hardness.

Occurrence: Obsidian is a super-cooled rock, occurring where lava has solidified rapidly, preventing the development of well-formed crystals. It is usually associated with rhyolitic eruptions and flows of viscous, acid lava.

▼ Snowflake obsidian from Utah, USA.

▶ Obsidian from Mývatn, Iceland, showing conchoidal fracture.

Pitchstone

Composition: Pitchstone is a glassy rock, of acid composition, with the mineralogy of granite and rhyolite. This mineralogy is difficult to detect owing to the microscopic grain size. The rock very similar to obsidian in composition, except that it is hydrated (it contains water). It is named after its dull, resinous or tarry surface lustre. Pitchstone is often dark greenish black or brown.

▲ Pitchstone from the Isle of Eigg, Inner Hebrides, Scotland, UK.

Grain size, texture and structure: This lava has a glassy texture, with no crystals obvious, even with a hand lens. Under the microscope, pitchstone is seen as a mass of glass and small anhedral crystals (grain size less than 0.5 mm or $\frac{1}{50}$ in). It may contain more crystalline material than obsidian and can exhibit flow banding. Phenocrysts, mainly of feldspar, pyroxene and quartz, are far more common than in obsidian.

Occurrence: As with obsidian, pitchstone forms by the rapid cooling of lava, often in flows. It sometimes occurs in dykes and other small intrusions where cooling has been very rapid.

Granodiorite

Composition: Granodiorite has a similar composition to granite, though plagioclase feldspar is more abundant than orthoclase. As well as quartz, the rock contains mica, hornblende and pyroxene, sometimes with small amounts of accessory ilmenite and magnetite.

Grain size, texture and structure: This is a coarse-grained rock, with component crystals of more than 5 mm ($\frac{1}{5}$ in) in diameter. The crystals tend to be anhedral and produce a compact interlocking groundmass. Though it is a rock of pale to medium colour, commonly slightly darker than granite, granodiorite can vary from pinkish to pale grey, depending on the mineral content.

Occurrence: This is one of the commonest plutonic igneous rocks. It forms in relatively large intrusions, including plutons and stocks.

◀ Granodiorite from the USA.

Diorite

Composition: Diorite has an intermediate composition with a total silica content between 55% and 65%. Essential minerals are plagioclase feldspar, hornblende and pyroxene, with little or no quartz. With increased quartz content, the rock may be classified as granodiorite. The mineral content gives the rock an overall dark colour with a speckled appearance, grey and dark green or black predominating.

Grain size, texture and structure: This is a coarse-grained igneous rock, with crystals more than 5 mm (⅕ in) in diameter. Though generally an equigranular rock, it can have a porphyritic texture, with phenocrysts of hornblende or feldspar. Xenoliths are common in diorite intrusions.

▶ Diorite from Saint Clement, Jersey, Channel Islands.

Occurrence: Diorite often forms isolated dykes, bosses and stocks. The rock may be found in association with both acid granite and basic gabbro and can merge into intrusions of these rocks.

Syenite

Composition: An uncommon intermediate rock with 55% to 65% total silica, syenite may contain up to 10% quartz, but commonly there is very little. As the quartz content increases, the rock can grade into granite. Predominant minerals are alkali feldspars, mainly orthoclase, with some sodium plagioclase, biotite mica, amphiboles and pyroxenes. Accessory minerals include titanite, zircon and apatite. When nepheline is common in the rock, the name nepheline syenite can be used. This mineralogy gives the rock a pale grey, white or reddish overall colour.

Grain size, texture and structure: Syenite is a coarse-grained rock, usually with an equigranular texture, though it may be porphyritic. Small cavities within the rock are relatively common. Pegmatitic varieties of syenite (with very coarse grain size) are sometimes found.

Occurrence: This igneous rock is very similar to granite, but less common. It forms in various intrusions such as stocks and dykes. Intrusions of syenite are rarely as large as those of granite, into which it can gradually grade.

▶ Nepheline syenite from Larvik, Norway.

Andesite

Composition: Andesite is an intermediate igneous rock containing a total of 55% to 65% silica. It has much plagioclase feldspar with pyroxenes, amphiboles and biotite mica. Andesite is a dark to medium coloured rock, being grey, greenish or brown. Mineralogically it is very similar to the coarse-grained rocks syenite and diorite.

▲ Andesite from Iztaccíhuatl, Mexico.

Grain size, texture and structure: This is a fine-grained rock with crystals less than 0.5 mm ($\frac{1}{50}$ in) in diameter. It may have glassy, very rapidly cooled components. Many andesites exhibit a porphyritic texture, with phenocrysts of feldspar, pyroxenes and amphiboles. These are set in the fine-grained groundmass of the rock as euhedral crystals. Similar minerals form the groundmass. Andesite can also have a vesicular or amygdaloidal texture, with cavities made by gas bubbles in the original lava being either empty or filled with amygdaloidal minerals, including zeolites. Flow structures are not unusual and xenoliths, rock fragments caught up in the original lava, may occur.

Occurrence: Andesite is a common solidified lava that forms in flows and other volcanic structures, as well as in small intrusions such as dykes. It often occurs with other types of lava, especially basalt, though andesite produces more explosive eruptions.

Trachyte

Composition: This is an intermediate rock with a total silica content of 55% to 65%. The feldspars in this rock are potassium- or sodium-rich alkali feldspars. There may be small amounts of quartz (less than 10%). As well as feldspar, trachyte is composed of pyroxenes and amphiboles with biotite mica. The composition is similar to that of coarse-grained syenite. The mineralogy gives an overall grey, whitish, yellowish or pinkish colouring.

▲ Trachyte from the Atlantic Ocean.

Grain size, texture and structure: Trachyte is a fine-grained volcanic rock, with the groundmass crystals smaller than 0.5 mm ($\frac{1}{50}$ in) in diameter. Most trachytes have a porphyritic texture, containing small euhedral phenocrysts, mainly of feldspar. Flow structure is common in trachyte, parallel lath-shaped feldspar crystals producing a 'trachytic' texture when they curve around the larger phenocrysts. This structure may be too small to observe with the naked eye.

Occurrence: Trachyte forms from cooled lava and can also occur in small intrusions such as dykes and sills. It is often found in association with the more extensive and common rock, basalt.

Gabbro

Composition: Gabbro is a basic rock with between 45% and 55% total silica. It is composed of plagioclase and pyroxene, sometimes with small amounts of quartz. Olivine can be a common mineral in gabbro, in which case the rock is called olivine gabbro. Hornblende is also not uncommon and accessory minerals include magnetite and chromite. The overall colour is speckled dark and light, with pale feldspar and darker, often black or greenish pyroxenes.

Grain size, texture and structure: Gabbro is coarse-grained, with crystals over 5 mm (⅕ in) in diameter. It is generally an equigranular rock, porphyritic texture being unusual. A common texture in gabbro occurs when single plagioclase crystals are surrounded by pyroxene (ophitic texture). A layered structure is frequently found, dark layers of rock alternating with paler layers. The dark layers are rich in pyroxene, while the paler layers have a predominance of feldspar. This structure can be on a small scale, a few centimetres per layer, or can occur over a large area, with layers over a metre in thickness. Layering is related to the intrusion of the gabbroic magma and its cooling history.

Occurrence: Gabbro occurs in dykes and sills, as well as in large plutonic stocks. It can form in sheet intrusions such as lopoliths.

▲ Gabbro from Silesia, Poland.

▶ Layered 'bird's eye' gabbro from St Peter Port, Guernsey, Channel Islands.

Dolerite (diabase)

Composition: Dolerite has a basic composition, containing between 45% and 55% total silica and less than 10% quartz. This composition is the same as the coarse-grained equivalent, gabbro. Dolerite is composed of two main minerals: plagioclase feldspar and pyroxene (usually augite). Olivine, hornblende, biotite and magnetite can be present in small amounts. If there is a significant amount of olivine, the term olivine dolerite may be used. This mineralogy gives dolerite a dark colour, often dark grey, black or greenish-black, weathering to a rusty brown.

▶ Dolerite from Scourie, Sutherland, Scotland, UK.

Grain size, texture and structure: A medium-grained rock, dolerite has component crystals between 0.5 mm (¹⁄₅₀ in) and 5 mm (¹⁄₅ in) in diameter. The texture is usually equigranular and the crystals euhedral, but microscopic magnification is really needed to see them. Phenocrysts of feldspar may give the rock a porphyritic texture. Sometimes an ophitic texture, with pyroxene surrounding plagioclase, can occur. Dolerite intrusions may exhibit columnar jointing, though this is rarely as well formed as in lava flows. In a vertical dyke this structure is horizontal, whereas in a sill the columns are vertical.

Occurrence: Dolerite is a common rock in small igneous intrusions such as dykes and sills. In many areas, such as the Inner Hebrides of western Scotland, swarms of dykes occur, forming much of the crust and radiating from volcanic centres. As dolerite is a very durable rock, these dykes are often observed as ridges across the landscape.

Basalt

Composition: A basic igneous rock with similar overall composition to gabbro and dolerite, basalt contains about 50% plagioclase (generally calcium-rich) and 50% pyroxene (commonly augite). Olivine can also be present (olivine basalt), as can magnetite and small amounts of quartz. Various minerals occur in vesicles, forming amygdales, and include zeolites, calcite, quartz and chalcedony, which can be in the form of concentrically banded agate. Basalt is a very dark coloured rock, often nearly black, though it can have a greenish tint.

Grain size, texture and structure: Basalt is a fine-grained rock and, even with a x10 hand lens, the component crystals can rarely be seen. The individual crystals are less than 0.5 mm (¹⁄₅₀ in) in size and may be anhedral or euhedral. Porphyritic texture is common in basalts, with phenocrysts of plagioclase, olivine and augite set in the fine-grained groundmass. Vesicular texture (with small, often rounded cavities) and amygdaloidal texture (with infilled vesicles) are both common. On a large scale, basalt often displays magnificent columnar jointing. The almost perfect columns are usually hexagonal (six-sided) and are related to the cooling of the molten lava from which

▶ Vesicular basalt, probably from Iceland.

▶ Ropy basalt from Hawaii, USA.

basalt forms. Because it is a very fluid lava, basalt flows can be extensive. A rounded structure formed by underwater eruptions is called pillow lava. This occurs when an outer skin solidifies on a mass of lava and the mobile, molten rock then fills in the rounded shape, before finally solidifying. Basaltic lava flows and masses can exhibit different surface forms. Flows can have a ropy surface structure called 'pahoehoe', or a blocky surface called 'aa'. These Hawaiian words are now part of geological language.

Occurrence: Basalt forms in lava flows, dykes and sills. Much of the Earth's crust is made of basalt, as it covers the ocean floor, often beneath a relatively thin veneer of sediment. Basaltic volcanoes tend to be low in height and very extensive, because of the fluid, often non-violent, free-flowing nature of the eruption. The Moon's surface is mainly composed of basalt.

Anorthosite

Composition: Anorthosite is a basic rock containing much plagioclase, possibly 90% or more. Other minerals include olivine, magnetite and pyroxene, but these are in small amounts and can be classified as accessories. The total silica content is less than 55%. Anorthosite is a relatively pale-coloured rock, generally grey or whitish.

Grain size, texture and structure: This is usually a coarse-grained rock, with crystals over 5 mm (⅕ in) in diameter, but can be medium-grained. The component crystals give anorthosite a granular texture with euhedral crystals. The rock may have bands of dark minerals, these bands exhibiting a parallel orientation of the feldspar crystals. There can be a layered structure, like that of gabbro.

▶ Anorthosite from Montana, USA.

Occurrence: A plutonic rock, anorthosite occurs in large batholiths and stocks and also forms dykes, commonly associated with gabbro intrusions. It can be found over vast areas and may, with an increase in pyroxene and other ferro-magnesian minerals, grade into gabbro. Anorthosite often occurs in areas of regionally metamorphosed rock.

Larvikite

Composition: Larvikite is an intermediate igneous rock with a composition very similar to that of syenite. Its total silica content is between 55% and 65%. It contains plagioclase feldspar, pyroxene and amphibole. The plagioclase is characterized by a blue-grey colouring and shows schillerization (an iridescent play of light), which gives the rock its typical appearance. Accessory minerals may include apatite and augite. Larvikite has an overall bluish-grey colour.

Grain size, texture and structure: This is a coarse-grained igneous rock, the component euhedral crystals being more than 5 mm (⅕ in) in size. The feldspar crystals may be over a centimetre in length. Pyroxenes and other mafic minerals can form patches within the rock groundmass.

▲ Larvikite from Vestfold, Norway.

Occurrence: Larvikite is an intrusive igneous rock that forms in small intrusions such as dykes and sills.

Peridotite

Composition: As an ultrabasic rock, peridotite contains less than 45% silica with virtually no feldspar. It is a dense, dark-coloured rock, rich in heavy ferro-magnesian minerals, especially olivine, pyroxene and hornblende. Chromite and garnet may also be present. When the rock has a relatively high proportion of reddish garnet, giving it an attractive mottled appearance, the term garnet peridotite is applied. Dunite is a dark-greenish- or brownish-coloured rock, which is composed almost entirely of olivine. The name peridotite is derived from peridot, the name of gem-quality olivine.

Grain size, texture and structure: This is a coarse- to medium-grained rock, with crystals commonly over 5 mm (⅕ in) in size. It is usually equigranular, but porphyritic examples sometimes occur. Garnet peridotite may have porphyritic crystals of red or blackish garnet, up to 10 mm (²/₅ in) in size, set into the generally greenish groundmass.

Occurrence: Peridotite is an igneous rock formed from magma generated in the upper part of the Earth's mantle (the region below the crust). This rock is usually associated with basic magma, occurring in dykes and other bodies often in or near large intrusions of gabbro, sometimes as layers at the base of the gabbro. Peridotite may develop when olivine, which crystallizes very early during the cooling of gabbroic magma, sinks down due to its high density, forming a distinct rock layer at the base of the magma mass. This rock can be found in diamond pipes and as xenoliths in basaltic lavas that originated at considerable depth in the mantle, possibly as deep as 100 km (62 miles). In high-grade metamorphic rocks, discrete masses composed of peridotite can occur.

▶ Peridotite from the Atlantic Ocean.

Dunite

Composition: Dunite is a rock of ultrabasic composition containing less than 45% total silica. This rock is composed almost entirely of the mineral olivine, giving it an overall brownish or green colouring. It can contain small amounts of pyroxene and chromite as accessory minerals. Dunite is a form of peridotite, and the alternative name olivinite is sometimes used.

▶ Dunite from Namibia, Africa.

Grain size, texture and structure: A coarse- to medium-grained rock, with crystals between 0.5 mm (¹⁄₅₀ in) and 5 mm (¹⁄₅ in) in size, dunite has an equigranular, sugary texture.

Occurrence: Dunite, as a type of peridotite, generally forms at deep levels in the Earth's crust, where ultrabasic rocks can result from the settling and differentiation of basic magma, creating smaller dykes and sills of ultrabasic composition. At times dunite occurs as small masses within gabbro intrusions, representing fragments brought from depth in the basic magma.

Tuff

Composition: This is a fragmental pyroclastic volcanic rock composed of a variety of materials, depending on the nature of the volcanic eruption. If the fragments ejected contain much crystalline material, the tuff is called crystal tuff. It may be composed of feldspar, quartz, pyroxene and amphibole. Tuff can also contain rock fragments (lithic tuff). The rock fragments are generally types of lava, such as rhyolite, andesite and trachyte. Vitric tuff contains glassy material and pumice.

▼ Tuff from the USA.

Grain size, texture and structure: Tuff is a fine-grained rock with many components usually less than 0.5 mm (¹⁄₅₀ in) in size (ash). Some fragments (lapilli) may be up to 5 mm (¹⁄₅ in). The rock can have a porphyritic texture, with some small euhedral crystals set in the vesicular groundmass. Tuff that's been deposited in water often have bedding structures similar to those found in sedimentary rocks.

Occurrence: An eruptive volcanic rock, tuff is formed from the consolidation of the smaller fragments ejected from a volcano. Volcanic dust and ash are the main components. The finer-grained material may be thrown high into the atmosphere and carried on wind currents. The deposits of tuff with the greatest thickness occur near to the volcanic crater. Layers of tuff can be very valuable stratigraphically, as an individual layer is deposited at a single, definite time, and may cover a wide area.

Pumice

Composition: This rock generally has an acid composition and is very similar to rhyolite and obsidian; however, both intermediate and basic lavas can form pumice. Typical rhyolitic pumice contains silicates such as feldspar, ferromagnesian minerals and glassy quartz. Pumice is usually a pale-coloured rock.

Grain size, texture and structure: Pumice is a very fine-grained rock, any crystals only visible with microscopic examination. It is a highly vesicular rock full of small cavities. This reflects the original molten rock's highly frothy nature. It can be an amygdaloidal rock when the vesicles (gas cavities) are infilled with minerals such as quartz and zeolites. Mainly because of this vesicular texture, pumice has a very low specific gravity, often less than 1.0, and typically floats on water. Depending on the flow of the lava, pumice may have orientation of the vesicles.

▶ Pumice from Vulcano, Aeolian Islands, Italy.

Occurrence: A pyroclastic volcanic rock, pumice forms when frothy, gas-filled lava erupts on land or in water. When such eruptions occur in the sea, fragments of pumice may float for some distance. Pumice is frequently found as the result of violent nuée ardente eruptions, where clouds of incandescent gas, filled with droplets of lava, flow at speed down the volcano's slopes. The rock formed by such eruptions, ignimbrite, contains much glassy pumice.

Ignimbrite

Composition: Ignimbrite is a form of volcanic tuff and generally contains much glass, with feldspars, feldspathoids and mica, often having a composition similar to acid rhyolite. It also contains small rock fragments and some well-formed larger crystals. The overall colour is pale, being grey or brown, turning a reddish colour when weathered. Fragments of sedimentary rock may be caught up in ignimbrite.

◀ Ignimbrite from Patararu, New Zealand.

Grain size, texture and structure: This rock is a poorly sorted mass of fragments, usually small in size, less than 5 mm (⅛ in), with a fine-grained groundmass, commonly made of fragments of glass. Flow-banding is common, with gas cavities being aligned, as is a eutaxitic texture, where glass fragments have a rounded shape related to gas bubbles in the lava. Ignimbrite is a welded tuff, often with a hard surface. On a larger scale, columnar jointing can occur and individual flows may show a graded structure, with coarser fragments being found at the base and finer ones towards the top. Porphyritic texture, with phenocrysts of feldspar, is not uncommon.

Occurrence: Ignimbrite is the deposit from an incandescent volcanic cloud of gas, molten lava droplets and rock fragments. Such eruptions are very explosive and the cloud of material flows rapidly down the volcanic slope. Fine tuff and ash are held in the gas cloud, while larger material is found in the base of the flow.

Volcanic bombs

Composition: The composition of a volcanic bomb depends on that of the lava erupted from a particular volcano. Andesitic and, much less frequently, basaltic volcanoes are the main source of volcanic bombs. Basaltic volcanoes are far less explosive and so bombs rarely result from them. Minerals can include quartz, feldspar, amphibole and pyroxene. Bombs are usually pale to dark brown or grey in colour.

Grain size, texture and structure: Bombs are classified as agglomerate, a term that includes ejected volcanic fragments more than 5 mm (⅕ in) in size. Volcanic bombs, which are ejected in a molten condition, have two main structures. As molten lava is carried through the air, it may take on an elongated shape, forming spindle bombs, as the lava twists during its fall. Breadcrust bombs have an irregular outline and are marked by numerous deep surface cracks. These cracks develop after consolidation of the surface, as the still-molten interior expands, though some cracks may form when the partly solidified clot of lava hits the ground. Such bombs can be over a metre (39 in) in size.

Occurrence: Volcanic bombs are ejected from erupting volcanoes as clots of lava. Larger fragments of agglomerate, including bombs, are often confined to the crater or sides of the volcano.

▶ Spherical basalt volcanic bomb from Kilauea, Hawaii, USA.

Metamorphic rocks

These rocks are formed when existing rocks are altered in a number of ways. Magma or lava can heat and change pre-existing rocks, producing recrystallised rocks (contact metamorphism). Pressure and temperature deep in the crust alter rocks (regional metamorphism) and large-scale thrust faults can pulverise and change rocks dynamically.

Slate

Composition: Slate is the result of low-grade regional metamorphism (low temperature and pressure conditions) of fine-grained sedimentary rocks such as shale, clay and volcanic tuff. Constituent minerals include quartz and micas, with chlorite, feldspars, graphite and minerals of the clay group. It can also include porphyroblasts of pyrite, which are seen as euhedral (well-formed), often cube-shaped, crystals in the rock groundmass. Slate varies in colour and may be black, grey or greenish, depending on the mineral content of the original rock.

Grain size, texture and structure: This rock is equigranular and fine-grained, apart from possible porphyroblasts. The grains in the groundmass tend to be anhedral but are only really visible with microscopic examination. Slate is characterized by its cleavage. This is a structure caused by metamorphism, which aligns the component particles parallel to each other and at right angles to the direction of metamorphic pressure. Slaty cleavage allows the rock to be readily split into thin sheets. This structure is not related to the bedding of the pre-metamorphosed rock. Original features, such as bedding and other sedimentary structures, can still be apparent, as may fossils, which are often distorted.

Occurrence: Slate occurs on the margins of mountain belts, where metamorphic conditions are not extreme.

▶ Slate with pyrite from Cumbria, England, UK.

Phyllite

Composition: Phyllite is composed of quartz and feldspars, with micas and clay minerals. Chlorite is a common component, giving the rock a greenish colour. Micas give a silvery sheen to the rock surfaces. Porphyroblasts of garnet may occur.

▶ Phyllite from Switzerland.

Grain size, texture and structure: This rock is medium- to fine-grained and equigranular, being derived, by regional metamorphism, from pelitic sedimentary rocks. The groundmass components tend to be anhedral, though some euhedral mineral crystals may be visible with a microscope and any garnet porphyroblasts are usually euhedral. Though not as easily cleaved as slate, phyllite has a more wavy, undulating structure related to metamorphic pressures and it can show folding on a small sale.

Occurrence: Phyllite is between slate and schist in metamorphic grade and is found in the outer regions of fold mountain belts, often grading into schist.

Schist

Composition: Schist is rich in micas, with quartz and feldspars making up much of the rock. Some minerals such as garnet, kyanite and biotite may be common enough for the rock to be named after them, for example kyanite schist. These minerals are the result of metamorphic conditions. The crystals in schist are both anhedral and euhedral, and the rock is often silvery because of its mica content. Schist can be various colours, including dark grey, greenish or brownish.

▶ Schist from Aberdeenshire, Scotland, UK.

Grain size, texture and structure: This a medium-grained rock, with most of the minerals visible to the naked eye and certainly with a x10 hand lens. The rock has an equigranular texture, often with porphyroblasts of garnet or staurolite. The structure of schist is characterized by a pronounced wavy banding (schistosity). Component minerals emphasize this structure; mica, for example, lies on the banding surfaces. Layers of different minerals such as feldspars, quartz, micas, hornblende and chlorite may run through the rock, giving it an alternating dark and light appearance. The rock is often folded, and in some schists relict sedimentary structures can be visible.

Occurrence: Schist is the product of medium-grade regional metamorphism, where temperatures and pressures are considerably higher than those that form slate. Most sedimentary rocks and some metamorphic and igneous rocks are altered. Schist is found nearer the centres of fold belts. Different minerals in the rock can indicate varying metamorphic conditions. Schists with abundant chlorite are formed at a lower grade of metamorphism than those containing biotite, while garnet-bearing schists are formed at a higher grade.

Gneiss

Composition: Gneiss contains much quartz, feldspars and micas, giving it a granitic composition. Other minerals include hornblende, pyroxenes and garnet. Gneiss is sometimes pale coloured, but there are usually separate bands of dark- and light-coloured minerals. Orthoclase feldspar in the rock may give it a pinkish colour. The pale bands are rich in feldspars and quartz, while the darker ones consist of biotite mica and hornblende.

Grain size, texture and structure: This is a coarse-grained rock, the component, generally euhedral, minerals being over 2 mm ($^1/_{12}$ in) in size and easily visible. The rock has an equigranular texture, though there are often porphyroblasts of minerals such as garnet. A type of gneiss called augen gneiss is characterized by large – over 1 cm (2/5 in) in size – rounded, pale-coloured porphyroblasts of quartz and feldspars. Darker

▶ Garnet-biotite-gneiss from Bellinzona, Ticino, Switzerland.

components such as biotite mica may swirl around these. The characteristic structure of gneiss is its banding, where dark and light minerals are separated into alternating bands in the rock.

Occurrence: Gneiss is formed at the highest grade of regional metamorphism and all types of rock can be altered. The temperature and pressure conditions creating this high-grade rock are extreme, occurring only at great depth in the Earth's crust and in the central part of fold mountain belts. Gneiss often forms in such extreme conditions that it is close to where melting occurs and it can grade into granite here.

Migmatite

Composition: This metamorphic rock is a mixture of dark material of basic composition and paler granitic rock. Each component has its own typical mineralogy: the dark rock may contain amphiboles, pyroxenes and biotite mica, the paler material being composed of feldspars, quartz and micas.

Grain size, texture and structure: Migmatite is a coarse-grained rock with a granular texture. The component minerals are easy to recognize, even without a hand lens, and crystals are generally euhedral. There may be a gneissose banding structure, or lenses and pods of granitic material, giving an appearance similar to augen gneiss. Larger porphyroblasts of minerals such as feldspars and quartz are often present. The folding frequently seen

in migmatite suggests that the rock has gone through an almost molten, plastic stage in its formation. At times, these folds are complex and highly distorted and are referred to as ptygmatic folds.

Occurrence: This rock, which may be a mixture of metamorphic and igneous material, forms very deep in high-grade metamorphic belts and also in the aureoles around granite intrusions. It is possible that granitic components were intruded into previously formed high-grade metamorphic rocks.

▶ Migmatite from Jihomoravský, Czech Republic.

Eclogite

Composition: Eclogite is composed of dense ferromagnesian minerals such as pyroxenes and garnet. This mineral content gives the rock a very similar chemistry to that of an ultrabasic igneous rock, as it is very low in total silica. The pyroxene is often a greenish variety called omphacite, and when it also contains red garnet, the rock has a striking appearance, being mottled with these two colours. It may contain hornblende and kyanite.

▶ Eclogite from Cape Province, South Africa.

Grain size, texture and structure: This is a medium- to coarse-grained, granular rock with the euhedral component garnets and pyroxenes occurring as crystals over 5 mm (¹/₅ in) in size. Larger porphyroblasts of these minerals can occur within the generally equigranular texture. All the crystals in eclogite are easily seen with the naked eye. It is a crystalline rock, and though often without structure, it may be banded.

Occurrence: Eclogite forms at the highest grade of metamorphism in the deepest levels of the Earth's crust, with peridotite and serpentinite. It can often be found as large masses in other metamorphic rocks, and this, together with its high density, is possible evidence that eclogite has been carried from the base of the crust or upper regions of the mantle. The rock may be the result of high-grade metamorphism of basic igneous rocks such as gabbro.

Amphibolite

Composition: The main minerals in this high-grade metamorphic rock are plagioclase feldspars and amphiboles, commonly hornblende, actinolite and tremolite, giving the rock a greenish colouring. Garnet, feldspars and pyroxenes can also occur. This mineralogy produces a composition similar to that of an ultrabasic igneous rock.

▶ Amphibolite from Val Arbedo, near Bellinzona, Ticino, Switzerland.

Grain size, texture and structure: Amphibolite is a medium- to coarse-grained rock, which is mainly equigranular, though there can be porphyroblasts of garnet and other ferromagnesian minerals. The rock may exhibit schistosity. The amphibole minerals in amphibolite frequently have a needle-like habit.

Occurrence: This rock is usually the result of high-grade metamorphism of igneous rocks such as dolerite. It often occurs as masses, with an intrusive appearance, among metamorphosed sedimentary rocks. Amphibolite is also found within gneiss and schist as lenses and pods.

Serpentinite

Composition: Serpentinite is composed mainly of serpentine group minerals such as chrysotile and antigorite. It also contains other ferromagnesian minerals including garnet, amphiboles and pyroxenes, as well as chromite and iron oxides. Relict olivine can be present, which gives the rock a dark colour, often greenish or black. Serpentinite has a low silica content.

▶ Serpentinite from Essex, Massachusetts, USA.

Grain size, texture and structure: This rock is coarse-grained and crystalline, with most of the minerals easily visible to the naked eye. Component crystals are generally euhedral, but there are commonly veins of finer, medium-grained material, giving a banded appearance.

Occurrence: Serpentinite forms where igneous rocks such as peridotite and basalt have been altered by serpentinization. This is a process involving metamorphic changes brought about by heating and the movement of high-temperature water-based fluids. It occurs at great depth in the Earth's crust, possibly deep below the ocean floor. Serpentinites are often found as discrete masses within areas of regionally metamorphosed rocks. Though they are altered, metamorphosed rocks, some scientists group serpentinites with the igneous rocks.

Marble

Composition: This rock is composed mainly of calcite and/or dolomite, as it is metamorphosed limestone. Any impurities in the original limestone may be altered to other minerals, so marble can contain minerals such as brucite, olivine, diopside, wollastonite, tremolite and serpentine minerals. These can result in coloured patches and veins within the overall pale-coloured rock. Brucite mottles the rock with greenish and blue colours.

◀ Marble from Mijas, Spain.

Grain size, texture and structure: Marble is a crystalline rock, with a fine- to coarse-grained texture and often a sugary appearance. It can have a granoblastic texture, with anhedral, interlocking crystals. When viewed with a microscope, or a hand lens, the interlocking calcite crystals are visible. Original sedimentary structures such as bedding may be apparent, depending on the degree of recrystallization, and pore spaces in the original limestone are usually absent.

Occurrence: Marble is produced by the metamorphism, either contact or regional, of limestone. It occurs within metamorphic aureoles, where it can be associated with hornfels. Metamorphic alteration is greatest near to an igneous intrusion and marble grades into the original limestone farther away from an igneous mass. When it is a product of regional metamorphism, it is associated with metaquartzite, schist and phyllite.

Hornfels

Composition: The composition of hornfels depends to a great extent on the mineralogy of the original non-metamorphosed rock and the type of magma producing the changes. Feldspars and quartz are dominant minerals, together with garnet, biotite mica, pyroxenes, cordierite and andalusite. Chiastolite (a type of andalusite) is common in some hornfels. The rock may be named after an important constituent mineral, for example, garnet hornfels, cordierite hornfels or pyroxene hornfels. Hornfels is a dark coloured rock, varying from grey to greenish.

Grain size, texture and structure: Hornfels has a fine- to medium-grained equigranular texture. The crystals are often too small to be recognized without a hand lens or microscope. It is a crystalline rock with a 'flinty' appearance. It may have a granoblastic texture, when larger crystals of garnet or andalusite are set in the rock groundmass. Chiastolite porphyroblasts show up as thin, pale, lath-shaped crystals, which are cross-shaped in section.

▼ Dark hornfels in contact with pale granite, from Cornwall, England, UK.

Occurrence: This is a rock formed by contact metamorphism of pelitic rocks, especially shales and mudstones. Near the igneous contact, virtually complete recrystallization occurs, while gradually the features of the original rock may be encountered farther away.

Spotted slate is very similar to hornfels but tends to be finer-grained and has cleavage planes, not unlike those in slate, formed by regional metamorphism. It is dark coloured, often almost black. The typical darker spots are usually small – less than 5 mm ($^1/_5$ in) – patches of andalusite or cordierite. Spotted slate occurs in the outer parts of metamorphic aureoles, and grades into hornfels closer to the magma body.

Metaquartzite

Composition: With over 90% quartz, this rock has a pale grey colour. It may contain small amounts of magnetite, micas and feldspars. Iron minerals can give the rock a pinkish or darker colouring.

Grain size, texture and structure: Metaquartzite is a crystalline rock with interlocking, anhedral quartz grains, very unlike the original sandstone, which contained many pore spaces. This fine- to medium-grained rock may have a sugary texture and is not unlike marble, but metaquartzite has a much harder surface and does not react to dilute hydrochloric acid. Original bedding structures can be visible.

▶ Metaquartzite from Whitehills, Scotland, UK.

Occurrence: Metaquartzite forms in contact aureoles near to a magma body, where heating is extreme. It occurs when sandstone has been altered by contact metamorphism.

Skarn

Composition: Skarn is a calcareous rock containing calcite and also various silicate minerals, including olivine, diopside, garnet, pyroxenes, serpentine minerals and wollastonite. Ore minerals, including galena, sphalerite, chalcopyrite and pyrite, can be associated with skarn. Molybdenum and manganese minerals in skarn may be of sufficient quantity to have economic significance. Skarn is a pale to dark coloured rock, with numerous patches of grey, brown, black and green.

▼ Skarn from Mt. Somma, Naples, Italy.

Grain size, texture and structure: This is usually a fine- to medium-grained rock, but some examples are coarse-grained, depending on the metamorphic conditions and the nature of the original rock. The crystals are generally euhedral, and may be concentrated into layers, nodules and zones within the rock.

Occurrence: Skarn is usually the result of contact metamorphism by granitic magma, though intermediate magmas can also create this rock. Impurities in the limestone from which skarn is formed, together with material from the magma, create the variety of minerals in skarn.

Mylonite

Composition: The mineralogy of mylonite is very variable, as it may be derived from a wide variety of rocks. It contains minerals that have formed during metamorphism in addition to pulverized 'rock flour' created by the metamorphic process. It can be dark or pale coloured, often greyish or brownish, with paler bands.

◀ Mylonite from Sutherland, Scotland, UK.

Grain size, texture and structure: Mylonite is a fine-grained rock, with anhedral components pulverized by the metamorphic process. There are often small lenses and streaks of different minerals, which can give the rock a slightly banded appearance. Foliations may be created where these bands have been bent and folded. Porphyroblasts and small patches of rock that have survived metamorphism are often present in the groundmass.

Occurrence: This rock occurs where large-scale thrust faulting has taken place. It is the result of rock being crushed as it is moved along near the thrust plane and is said to be the product of dynamic metamorphism. Shearing stress is intense in such situations, though temperature is not very significant. Mylonite is associated with mountain building and the relative movement of rock masses.

Sedimentary rocks

Sedimentary rocks develop on or near the Earth's surface. Three main categories are recognised: detrital rocks are composed of fragments derived from other rocks, organic rocks contain biologically formed material and chemical sediments result from inorganic chemical processes. Most sedimentary rocks are readily recognised by their bedding planes and many contain fossils.

Conglomerate

Composition: This rock can contain a variety of materials. It may be composed of different rock fragments, frequently together with quartz particles and other minerals that resist erosion and weathering. The components of conglomerates are generally related to the rocks in the area from which the fragments are derived. These particles are commonly held in a groundmass of sand, iron oxides, quartz or calcite. A conglomerate can be named after its content, for example, quartz conglomerate, or, when it contains a wide variety of rock fragments, polygenetic conglomerate.

Grain size, texture and structure: Conglomerate is an extremely coarse-grained rock, with most of the particles well over 5 mm (⅕ in) in size. There may be many pebbles and boulder-sized fragments set in a much finer-grained groundmass. The larger fragments have been rounded by the action of water during erosion and transport from the source area. Bedding structures are rarely apparent, though there can be alignment and grading of the particles. Such structures are usually only seen in the field, and fossils are generally lacking, apart from any within the rock fragments.

Occurrence: Conglomerate occurs in a variety of geological situations. It is associated with high energy environments, where strong water currents have moved and shaped the large fragments. It forms along shorelines and in shallow rivers. Conglomerate often occurs immediately above an unconformity. This is an old eroded land surface, on which the 'basal' conglomerate is the first layer of a new series of sedimentary rocks. In this situation, the conglomerate may be a beach deposit, formed as the sea level was rising.

▶ Hertfordshire pudding stone, a conglomerate from Pinner, Middlesex, England, UK.

Breccia

Composition: As with conglomerate, breccia can contain a wide variety of fragments derived from different rocks or may be composed of resistant materials such as quartz. These fragments are generally held in a finer groundmass of silt, sand, quartz or calcite.

▲ Breccia from Oran, Algeria.

Grain size, texture and structure: Breccia is a coarse-grained rock, composed of very large – over 5 mm (¹/₅ in) – angular fragments (contrasting with the rounded fragments in conglomerate) in a finer groundmass. It is a poorly sorted rock. The large fragments can vary greatly in size and they may have a random orientation. Bedding is rarely seen in hand specimens, but in the field it is sometimes apparent in the groundmass.

Occurrence: Breccia usually forms where mechanical weathering is active. It is normally a rock of dry conditions, often developing as a scree deposit on hillsides, commonly at the base of a cliff. When faulting occurs, a fault breccia can form along the fault line, where rocks have been fragmented by rock movement.

Sandstone

Composition: The term sandstone covers rocks of varying composition, but essentially sandstone is a rock containing much quartz, frequently with feldspars and micas. When the rock has around 25% feldspar, it is called arkose. This is grey, pink or red in colour. Small flakes of mica are often visible on the bedding planes. Glauconite is present in some sandstones and gives the rock a greenish colour (greensand). Red sandstones are coloured by iron oxides such as hematite, which may accumulate on and between the quartz grains. Some sandstones contain calcite as a cement. Sandstones with much quartz are said to be mature sedimentary rocks, as the particles have been through considerable erosion and weathering – processes that remove less resistant material.

Grain size, texture and structure: Sandstone is a medium- to fine-grained rock that is generally well-sorted, containing fragments of much the same size. The grains may be angular, as in gritstone, when the particles have been eroded and deposited by water. Other sandstones can have rounded, 'millet seed' grains and are usually associated with wind deposition. Such sediments are typically formed in arid regions. A wide variety of sedimentary structures are common in sandstone. These include bedding planes, cross-bedding, caused by sedimentation in a moving current of water or wind, and ripple marks. Such structures help geologists to indicate the environment of deposition. Some sandstones contain fossils. Most are porous rocks, which can hold liquids in the spaces between the grains, creating aquifers.

Occurrence: Sandstone forms in a variety of geological situations. Sands containing angular fragments tend to occur in marine or river environments, while those with rounded grains are found in desert regions. Arkose, an immature sandstone with an abundance of feldspar, may form near to the source area and have rapid deposition, as feldspar is readily weathered. This rock is generally derived from granite or gneiss.

▼ Arkose from Lapoirie near Remiremont, Vosges, France.

▲ Red sandstone from Cumbria, England, UK.

▼ Millstone grit (gritstone), from Derbyshire, England, UK.

▲ Millet seed sandstone from Penrith, Cumbria, England, UK.

Greywacke

Composition: Greywacke consists of a variety of minerals, mainly quartz and feldspars, together with chlorite and rock fragments. It is a dark coloured rock, often grey or greenish.

Grain size, texture and structure: This rock is a coarse type of sandstone, containing many grains over 2 mm ($^{1}/_{12}$ in) in size. Greywacke is poorly sorted, with grain sizes varying widely, the groundmass being clayey and much finer-grained than the fragments of quartz and feldspar. The components are mostly angular. Sedimentary structures are common in greywacke, especially graded bedding, which is best seen in the field. The base of a bed may be coarser than the upper parts, grading gradually upwards into clay-size fragments. Other structures include sole marks and slumps.

Occurrence: Greywacke can be deposited by a turbidity current, a rapidly flowing sea-

▼ Greywacke from Chile.

bed current carrying sediment of various grain sizes. These currents flow down the continental slope, taking sediment into deep water. After these have been deposited on the sea bed, finer material settles on top, with the final deposition consisting of very fine-grained sediment that has been held in suspension in the current. Greywacke can also form in other environments, including deltas and river floodplains.

Orthoquartzite

Composition: Orthoquartzite is a sandstone composed almost entirely of quartz fragments cemented by quartz. The rock commonly contains over 95% quartz. Rarely, small amounts of feldspars and rock fragments are found. This composition creates a pale, almost white rock, though grey and pinkish orthoquartzites also occur.

Grain size, texture and structure: A medium-grained, well-sorted sandstone, with grains held by a quartz cement, producing a hard rock. Orthoquartzite may contain various sedimentary structures, including stratification. Cross-bedding is not uncommon, and ripple marks sometimes occur. Fossils are rare.

▼ Orthoquartzite from Finger Mountain, Western Mountains, Antarctica.

Occurrence: Orthoquartzite is a 'pure' quartz sandstone with a mature composition. This suggests it forms over a long period of weathering and erosion, processes that have removed the minerals more prone to decay. It is generally deposited in marine environments.

Shale/mudstone

Composition: These similar rocks are composed of quartz, micas and feldspars with clay minerals. Other constituents can include pyrite, calcite, iron oxides and dark carbon-rich material. Pyrite may occur as distinct crystals on shale bedding planes. Shale and mudstone are dark-coloured rocks, commonly black, dark grey, dark green, brownish or reddish.

Grain size, texture and structure: Shale and mudstone are composed of very small particles, less than 0.005 mm in size, which cannot be seen with the naked eye and may even be difficult to see through a microscope. The rocks are generally well-sorted, the grains being much the same size. The main difference between shale and mudstone is a structural one. Shale is finely bedded, or laminated, and often splits easily into layers. Mudstone lacks such obvious structure. Both sediments frequently contain fossils, which may be well-preserved in iron- or calcite-rich concretions or on bedding planes.

Occurrence: Shale and mudstone are compacted clay. They are usually of marine origin and their fossil content is often palaeoecologically useful. As they are such fine-grained rocks, the sediment from which they are formed may have been carried on currents into deep marine water. Shale can be found in sequences with limestone, sandstone and coal.

▲ Mudstone from Meux's Brewery Well, Tottenham Court Road, London, England, UK.

▲ Shale with print of fossil ammonite, form Whitby, North Yorkshire, England, UK.

Siltstone

Composition: Siltstone contains much quartz with micas (readily seen on bedding planes) and feldspars. The rock often contains organic material, together with calcite, which can act as a cement. There are fewer clay minerals than in shale and mudstone. Siltstone is generally dark coloured or may be black, greyish and brown or, when limonite is present, a yellowish colour.

Grain size, texture and structure: This rock is slightly coarser-grained than shale and mudstone, but the individual anhedral grains are difficult to distinguish with the naked eye, a x10 hand lens generally being needed. Siltstone frequently exhibits bedding and other structures such as cross-bedding and ripple marks. Variations in the grain size and mineralogy may emphasize the laminations. Siltstone can be very fossiliferous. Calcareous or iron-rich nodules often occur, usually aligned along bedding planes.

Occurrence: Siltstone results from the compaction of silt-grade sediment that may accumulate in many different environments. Its grain size is between that of shale and fine-grained sandstone. Any fossil content and sedimentary structures in the rock can help to determine its origin, which may be either marine or freshwater.

▼ Banded siltstone from North Yorkshire, England, UK.

Clay/marl

Composition: Clay contains a high percentage of clay minerals, usually kaolinite, illite and montmorillonite. These minerals are often formed by the alteration of feldspars during chemical weathering or hydrothermal activity. There can also be quartz, feldspars and micas, with iron minerals such as hematite and limonite, which give the rock a reddish or yellowish colouring, respectively, though generally clay is a pale greyish colour. Calcite can be present. When the percentage of calcite is relatively high (40–60%), the rock is called marl. This calcareous rock is classified between limestone and clay. If it contains glauconite, marl can be a greenish colour, while red marl is coloured by iron oxides.

Grain size, texture and structure: These are very fine-grained rocks, with particles less than 0.004 mm in size, which cannot be seen with the naked eye and only with difficulty through a microscope. Clays often have an earthy appearance and lack bedding structures. Boulder clay is a sediment associated with glacial conditions. It consists of a very fine-grained rock flour of clay grade and contains glacially derived rock fragments, which are often angular.

Occurrence: Different clays occur in a variety of environments. Marine clays, which frequently contain fossils, may form in deep water. Some clays are lake deposits, while others are associated with glaciation. Varved clays, which have alternating seasonally formed dark and pale layers, are deposited in glacial lakes. The paler layers, which are thicker, represent the slightly warmer months of the year, when more sediment is available. The thinner, darker layers are formed in the cold season. China clay is the product of hydrothermal alteration of feldspars in granite.

▲ Clay from Harbury,
Warwickshire, England, UK.

Rock salt

Composition: Rock salt is mainly composed of the mineral halite, often as a mosaic of interlocking crystals. There is also detrital material in the rock, including clay minerals, quartz and silt. Impurities often cause colour variations. Rock salt can be a pinkish or red colour due to the presence of hematite (iron oxide). Silt makes the rock a greyish colour, but pure rock salt is colourless or white.

▲ Rock salt or halite.

Grain size, texture and structure: This crystalline rock usually has a massive (structureless) appearance and may have a sugary texture. Cubic halite crystals can often be seen. Bedding is generally absent, but when present there may be a distorted appearance, because rock salt can flow under pressure.

Occurrence: Rock salt is commonly found in sequences with other evaporites such as gypsum and polyhalite, interleaved with layers of shale and marl. Evaporites form when marine lagoons dry out and the salts contained in the water crystallize. This process can also occur in inland lakes. As the weight of overlying sediments may be considerable, layers of rock salt flow upwards to form large plug-shaped masses (salt domes), which break through the overlying strata.

Rock gypsum

Composition: This rock is formed of crystalline gypsum. Its colour usually varies from grey to white, but can often be reddish when stained by iron oxide impurities.

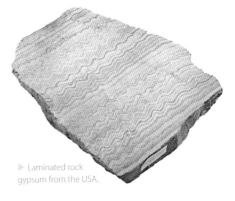

▶ Laminated rock gypsum from the USA.

Grain size, texture and structure: Rock gypsum is a bedded form of gypsum with fused crystals. It can have a sugary, earthy or fibrous texture and bedding structures that may be distorted. Rock gypsum beds sometimes have selenite crystals on their surfaces.

Occurrence: This is an evaporite rock, which is formed when marine lagoons and inland lakes dry out. Rock gypsum occurs with a variety of other evaporites, including rock salt, sylvite, anhydrite and polyhalite, in a sequence that can also include beds of shale, dolomite rock and marl. Evaporites are deposited in order of solubility. The least soluble, such as gypsum, are deposited first, and the most soluble, often halite, towards the end of crystallization. Gypsum rock also forms through the chemical alteration of anhydrite by the addition of water.

Oolitic limestone

Composition: This rock is essentially composed of calcite, with small amounts of detrital material such as quartz and silt. There are usually fossil fragments, or even complete fossilized organisms, in the rock. Oolitic limestone is a pale-coloured rock, commonly white or pale yellowish-brown. Iron impurities may give it a reddish colour.

Grain size, texture and structure: A medium- to coarse-grained rock, the individual grains (ooliths) are up to 2 mm ($1/12$ in) in diameter and can sometimes be seen with the naked eye. When the ooliths are pea-sized, the term pisolitic limestone is used. The individual ooliths are rounded or ovoid structures, which are built up in concentric layers around a sand grain or calcareous fossil fragment. The ooliths are held in a calcite cement. Cross-bedding is not uncommon in oolitic limestones.

▼ Oolitic limestone from Ketton, Rutland, England, UK.

Occurrence: In present times, oolitic limestone is being deposited in shallow, warm seas. Ooliths form by the accretion of calcite layers around small nuclei. It is thought that the constant movement and agitation of the water by tides and currents promotes this calcite precipitation. Fossil content often indicates marine conditions and cross-bedding suggests deposition in moving water.

Biogenic limestone

Composition: Within this category there are various common limestones, including coral, crinoidal and shelly limestone, named after the main fossils and fossil fragments they contain. Essentially these rocks are made of calcite, with small amounts of detrital sediment including quartz and clay minerals. The fossil fragments are held in a cement of hardened calcareous mud. Biogenic limestones may also contain quartz in the form of chert, which occurs as cryptocrystalline nodular masses, often in layers and bands following the rock's bedding planes. These rocks are usually pale grey or white, but can be pale brownish, when iron impurities are present.

Grain size, texture and structure: Biogenic limestones are coarsely crystalline rocks, commonly with a sugary groundmass of medium to large calcite crystals, and containing fossil fragments often over 5 mm ($1/5$ in) in size. Bedding is usually apparent in the field, and these limestones may show reef structures.

Occurrence: Biogenic limestones are usually of marine origin, as their fossil content can demonstrate. They form by the concentration of animal skeletons, including corals, crinoid fragments, brachiopods, molluscs and bryozoans. These rocks can be deposited in shallow conditions and some, containing fossils of free-swimming and floating creatures, may originate in deeper water. Freshwater biogenic limestones can be identified by the fossils they contain, such as the freshwater mollusc, *Planorbis*.

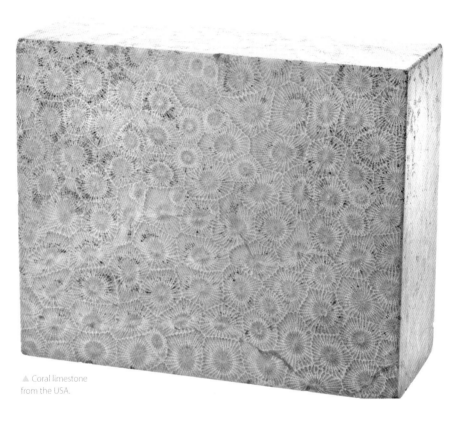

▲ Coral limestone
from the USA.

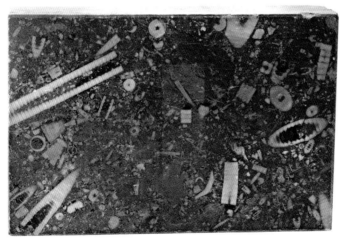

◀ Crinoidal limestone
from Derbyshire,
England, UK.

Dolomitic limestone/dolostone

Composition: The term 'dolostone' is often used for this rock instead of the former name 'dolomite', to avoid confusion with the mineral of the same name. Dolomitic limestone contains much magnesium-rich dolomite (magnesian limestone is another name for the rock), as well as a lower percentage of calcite. Also in the rock are detrital materials such as quartz and clay. Dolostone is a pale-coloured rock similar to biogenic limestone, but generally more cream or brown.

Grain size, texture and structure: This is a coarse- to fine-grained rock, which is usually crystalline, with a mass of dolomite crystals held in a lime-rich groundmass. Bedding structures may be present, but dolostone often has a massive appearance , lacking obvious bedding. The rock can contain fossils, though not in such quantities as in biogenic limestone due to recrystallization and the replacement of calcite and aragonite by dolomite. Vertical jointing, concretions and nodular masses are sometimes present and reef structures frequently occur.

Occurrence: Dolostones are often regarded as limestones that have undergone recrystallization. They occur with more calcite-rich limestones, and the recrystallization may take place when fluids migrating through the rock cause secondary replacement of minerals. They are usually formed in marine conditions. Some dolostones are associated with evaporite deposits containing rock salt, gypsum and anhydrite.

▼ Dolostone from Durham, England, UK.

Chalk

Composition: Chalk is an extremely pure form of limestone and is composed of a high percentage of calcite, with virtually no detrital material (clay or silt). The calcite in chalk is generally in the form of the remains of marine micro-organisms, including foraminiferans and coccoliths. Being such a pure calcite rock, chalk is very pale coloured, often white. However, it becomes darker and greyish as the amount of detrital material increases. If there is much iron oxide in the rock, it is known as red chalk.

▲ Chalk containing a flint nodule, from Devon, England, UK.

Grain size, texture and structure: This is a fine-grained rock, the component grains being examined best with a microscope. The rock can contain macro-fossils, which are set in the fine-grained groundmass. In a hand specimen, chalk usually appears massive, without structure, but bedding may be seen in the field. This is often emphasized by layers of silica-rich flint nodules. Underground layers of porous chalk can hold water and act as aquifers.

Occurrence: Chalk is a marine sedimentary rock that forms where no, or very little, land-derived sediment is being deposited. This may be because any land areas that could be a source of sediment were low lying and not subject to active erosion. The micro-organisms of which it is composed, and any macro-fossils, are of marine origin.

Travertine/tufa

Composition: Both travertine and tufa are inorganic limestones composed virtually entirely of calcite, though they may contain aragonite and small amounts of detrital material such as clay and quartz. Plant and other organic material are sometimes incorporated. They are white or cream rocks when pure, but yellowish, pale brown and rusty or reddish when iron oxide impurities are present.

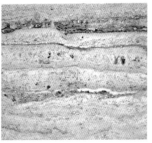

◀ Travertine from Italy.

Grain size, texture and structure: These are crystalline rocks, made of small to large crystals, which are generally fused together, giving a compact texture. Travertine is noted for its banded structure, whereas tufa is porous and spongy, without banding.

Occurrence: Travertine forms by the deposition of calcium carbonate from geothermally heated mineral-rich water, especially in terrestrial pools, hot springs and around volcanic vents. Tufa is deposited at much lower ambient temperatures than travertine. It is commonly found in cave systems such as in the limestone caves in northern England, and as coatings on rock surfaces and vegetation.

Ironstone

Composition: Ironstones vary considerably, but are all rich in iron-bearing minerals, including hematite, magnetite, siderite, goethite and limonite. Iron content is at least 15% of the rock and there can be much detrital material, including quartz and clay. Calcite is common in certain ironstones, often as a cement. Ironstones are usually dark brown, reddish or black. A high proportion of limonite produces a yellowish colouring.

Grain size, texture and structure: These rocks vary from fine- to coarse- grained. Oolitic ironstones have small rounded grains, as in oolitic limestone, but here they are formed from concentric layers of iron minerals, rather than calcite. Detrital grains tend to be angular. Bedding is common in many ironstones, and cross-bedding may be present. Banded ironstones have repeated layers of chert and siderite- or hematite-rich layers. This gives the rock a series of red and dark grey bands.

Occurrence: Many ironstones are produced by the alteration of primary sedimentary rocks by

▼ Oolitic ironstone from Radnitz mine, Plzeňský, Czech Republic.

the precipitation of iron minerals from solution. Replacement of calcite in limestone by iron-bearing solutions can be seen where certain fossils such as corals, originally made of calcite, have been converted to hematite. Oolitic ironstones may have originated as calcite-rich rocks in a marine environment and subsequently altered by iron-bearing fluids. Most banded ironstones occur in rocks of Pre-Cambrian age, possibly formed in non-marine basins. However, these rocks could have been deposited in mudflats and shallow marine situations. Early development of oxygen in the atmosphere may have been important in their formation.

Bauxite

Composition: Bauxite is an accumulation of a number of minerals; though it has often been classified as a mineral, it is better defined as a rock. The minerals in bauxite are mainly hydroxides of aluminium and iron. These include gibbsite, diaspore, bohmite, and limonite. Bauxite is reddish in colour and may contain darker patches. When it has a high limonite content, the rock is yellowish.

Grain size, texture and structure: A fine- to medium-grained rock, bauxite is usually massive, without bedding structures. Some bauxite deposits have a concretionary structure, while others are oolitic, pisolitic or earthy.

▶ Pisolitic bauxite from Kwahu West, Ghana.

Occurrence: Bauxite occurs in tropical regions, where weathering in humid conditions alters pre-formed rocks. Silicate minerals containing aluminium are changed by leaching; silica is removed and aluminium-rich, bauxite minerals are formed.

Coal

Composition: Coal is made of carbonaceous plant material and a number of volatiles, the percentage of which determine its quality as a fuel. There can also be various detrital particles, including silt and clay. High-quality coal (anthracite) is black in colour, with a vitreous lustre, but poorer coal is less lustrous (bituminous coal) and dirty to handle. Lignite is brown in colour and contains much visible plant matter. Jet is a type of coal that is valued for its high sheen, when polished, and black colouring. Jet has a brown streak when rubbed against a hard surface, whereas other coals produce a black powder.

Grain size, texture and structure: These rocks have a non-crystalline appearance, often with bedding structures when viewed in the field. Anthracite may be brittle and relatively hard, sometimes with a conchoidal fracture. Bituminous coal can show bedding and jointing. Lignite is less dense than higher quality coals and crumbles easily.

Occurrence: Coal forms by the compaction of thick peat deposits. The accumulation of sufficient peat in a damp environment generally requires anaerobic conditions. The weight of accumulated sediment above the initially deposited peat produces heat, which drives off volatiles, including water, increasing the relative carbon content and producing coal. The term 'rank' has been applied to the percentage of carbon in coal. Anthracite has a very high rank, with over 90% carbon. Bituminous coal contains more volatile material than anthracite and about 80% carbon. Coal tends to form in distinct layers called seams. These occur in cycles with shale, siltstone, sandstone and, at times, limestone. The repeated cycles of sediments associated with coal seams are called cyclothems. The great quantities of coal of Carboniferous age that provided fuel for the Industrial Revolution and are still used for electricity production in many countries, were formed on a vast delta system, on which prolific plant growth provided a considerable thickness of peat. Jet, which may contain fragments of plant material, forms from drifted wood, usually of the *Araucaria* tree, which subsequently sank to the sea bed and was covered in sediment. Jet occurs in marine shales as discrete layers and small masses.

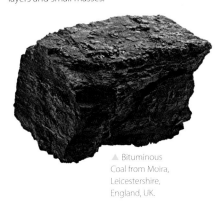

▲ Bituminous Coal from Moira, Leicestershire, England, UK.

◀ Lignite from Portrush, Northern Ireland, UK.

Amber

Composition: This is an organic substance, composed of hardened plant resin. It is relatively soft, easily worked and valued as a gemstone. Amber has a vitreous lustre and is transparent to translucent. During burial under sediments, its volatile content is reduced, increasing its chemical stability. Amber is usually a brownish or orange colour and frequently contains fossilized plant fragments and insects that were caught up in the original resin.

Grain size, texture and structure: Amber shows no individual grains and has an amorphous texture. When broken, it may produce a curved, conchoidal fracture.

Occurrence: This rock formed by the solidification and burial of resin produced by coniferous trees.

▼ Amber from Sicily.

Chert/flint

Composition: Both these rocks are made of silica, often in the form of chalcedony. Flint is usually regarded as a form of chert. Chert and flint vary in colour, ranging from pale grey to black. Flint may have a powdery, whitish surface and contain fossils; fossil echinoids (sea urchins) are often replaced by flint.

Grain size, texture and structure: The silica in the rocks is microcrystalline or cryptocrystalline and may be of organic origin. Chert and flint are very hard. Flint breaks with a curved, conchoidal fracture, leaving very sharp edges. There are sometimes cavities in the silica, which can contain earthy material, fossils or botryoidal chalcedony.

▼ Flint from Rableyheath, Hertfordshire, England, UK.

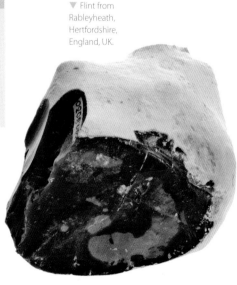

Occurrence: Although chert is often found as nodules and concretions in limestone, it can also form in beds. It may be produced by the accumulation of organic silica on the sea bed, by inorganic precipitation, or by the replacement of existing limestone. Flint occurs as rounded nodular masses in chalk, frequently following bedding planes and showing up as dark bands. It is thought that these nodules form from organic silica, which may become concentrated in sea-floor hollows and animal burrows, giving the silica shape.

Septarian concretion

Composition: Concretions usually have a similar composition to the rock in which they occur, and can be calcareous, siliceous or iron-rich. A concretion is harder than the surrounding sedimentary rock (often shale, mudstone or clay) and so may stand out from it. Concretions are similar in colour to the host rock.

Grain size, texture and structure: These structures are generally massive, without obvious grains, and form as rounded nodules of varying size, from a few centimetres to over a metre in diameter. Internally, septarian concretions are divided by radiating cracks with interconnecting veins, usually infilled with minerals such as calcite. The outer shell of a concretion may give no indication of the internal structure.

Occurrence: Septarian and other concretions and nodules occur in a variety of usually marine sedimentary rocks, especially shales and mudstones. They form during the conversion of soft sediment to rock during diagenesis, by the local segregation of minerals. A concretion may develop around a grain of sediment or organic remains and can contain three-dimensional fossils. Following formation, internal cracks are produced by shrinkage and these are infilled by minerals.

◄ Septarian-nodule.

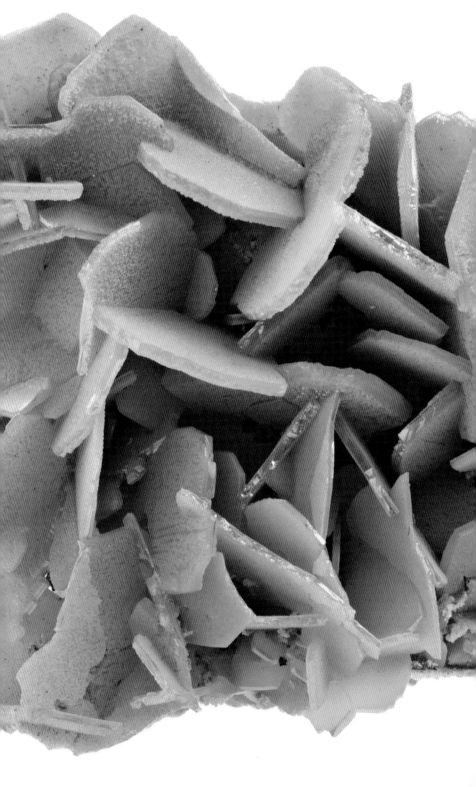

Introduction to minerals

Minerals are mainly inorganic solids, all formed naturally through geological processes. Most are chemical compounds, though a few, such as copper and sulphur, are single elements. Each mineral can be defined by its atomic structure and chemistry and many are identified by reference to certain physical or chemical properties, some of which are readily tested or observed. In order to identify a mineral specimen, it is necessary to work through these properties, build up an overall description of the mineral and compare this with the information in an identification reference book, such as this. It is important to realize that just one property alone will only very rarely provide a correct identification of a mineral specimen. There are over 5,500 different minerals and more are discovered each year.

THE OCCURRENCE OF MINERALS

The rocks that make up the Earth's crust are composed of minerals. Most rocks contain only a few different minerals; the igneous rock granite, for example, is made up mostly of feldspars, quartz and micas. Metamorphic and sedimentary rocks also consist of minerals, which may have originally formed in the rock or, as is often the case with sedimentary rocks, may be fragments derived from other rocks. The minerals of which rocks are composed are very rarely fine crystallized specimens; these generally occur where there are breaks in the Earth's crustal rocks, often along joints or faults. Hot fluids containing mineral-forming chemicals rise from depth into these fractures and, as the pressure is lowered, minerals are produced by deposition from the hydrothermal fluids. Such minerals are able to grow, often as terminated crystals, in the relatively unconfined spaces in the fractured rock, creating mineral veins. Metallic minerals including galena (an ore of lead) and sphalerite

(an ore of zinc) in these veins can be economically valuable and are frequently accompanied by uneconomic 'gangue' minerals such as quartz or barite. Mineral specimens with well-formed crystals can also develop in cavities in lavas and pegmatites. In lavas, these cavities (vesicles) are where bubbles of gas were trapped. Mineralizing fluids, similar to the hydrothermal vein fluids, seep into these vesicles and fine crystals, often of zeolites, calcite and amethyst, can grow in them. When not filled with crystals, many cavities are lined with concentric layers of agate. Crystals of evaporite minerals, such as halite and gypsum, are produced when saline water, rich in dissolved chemicals, dries out in lagoons or inland lakes.

THE SHAPE OF MINERALS

Minerals frequently occur in irregular shapes, especially in rocks with no cavities or fractures, as one mineral crystal is restricted by the growth of another. However, when minerals are able to develop freely, they can form perfect crystals, the shape of which is determined by the way their atoms are arranged. These crystals are classified into a number of groups according to their symmetry. Simple, everyday objects such as

◀ Wulfenite from Helena Mine, Mežica, Carinthia, Slovenia.

a matchbox can illustrate concepts of symmetry including planes and axes. If the box is cut in half parallel to any of the sides, the two halves are a mirror image of each other. The cut surface is a plane of symmetry. An axis of symmetry is an imaginary straight line, passing through the centre of one of the sides of the box and out through the centre of the opposite side. Rotation around this axis would produce the same view at least once. Though some crystals have a more complex shape than a matchbox, they can be interpreted in a similar manner.

There are seven crystal systems, but two have very similar symmetry and are often grouped together. The systems are defined according to their axes of symmetry.

The cubic (or isometric) crystal system is the most symmetrical, with three axes of equal length that intersect at right angles. There can be cubes, octahedra (eight-sided shapes), dodecahedra (twelve-sided shapes) and many combinations of these.

The tetragonal system comprises shapes that have three axes at right angles, but one is a different length from the other two. Crystals are prismatic, with a square cross-section.

The orthorhombic system has a lower degree of symmetry than the previous two, with three axes of differing lengths, all at right angles. Prismatic and tabular shapes are common in this system.

▼ Fluorite showing cubic crystals.

▲ Vesuvianite crystals illustrating the tetragonal system.

◀ Prismatic orthorhombic topaz crystal.

The monoclinic system is defined around three unequal axes, with two not at right angles to one another. The third axis is at right angles to the plane formed by the other two. This system also often contains prismatic and tabular shapes.

Crystals in the triclinic system have three axes of varying length and none is at right angles to the others. It has the lowest symmetry of all the groups.

The hexagonal and trigonal systems are often grouped together, as they are similar. They each have four axes of symmetry, three horizontal ones that are equal in length and a fourth vertical axis at right angles to these and of varying length. Crystals in these systems can be, but are not restricted to, six-sided prisms.

Twinning is found in many mineral specimens. This is when two or more crystals are joined in a way that follows specific rules of symmetry. Two common forms of twinning are contact twinning and interpenetrant twinning.

▶ Gypsum crystal, variety selenite, from Chihuahua, Mexico illustrating the monoclinic system.

▼ Thin, bladed crystals or plates of albite.

▶ Orthoclase crystals from near St. Gotthard, Switzerland.

◀ ▶ Tourmaline showing the trigonal system.

▶ Cerussite from Tsumeb Mine, Tsumeb, Namibia, illustrating twinning.

Probably more important from a practical identification point of view than the crystal system is the mineral habit. The habit of a mineral is the actual shape exhibited by a given specimen. This is characteristic in many common minerals and can be a good aid to identification. A variety of scientific and ordinary terms are used to describe mineral habits - some describe crystal shapes and others non-crystalline shapes. The following terms refer to important mineral habits:

Prismatic: of constant cross-section
Acicular: slender, needle-shaped
Bladed: elongated and thin, like a knife blade
Tabular: like a table top
Dendritic: tree-like in shape
Botryoidal: with the appearance of a bunch of grapes
Reniform: kidney-shaped
Mammillated: similar to botryoidal, with rounded shapes on a larger scale
Massive: irregular, with no definite shape

▼ Kyanite showing bladed crystals.

▼ Beryl crystal showing prismatic habit.

▲ Mass of Actinolite crystals, from South Dakota, USA showing radiating acicular habit.

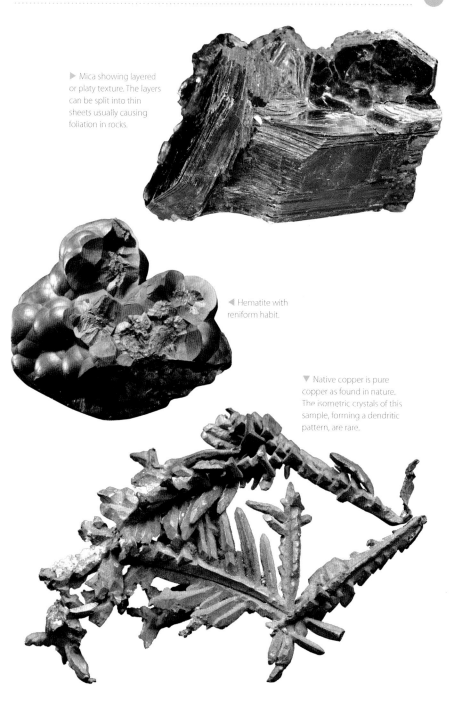

▶ Mica showing layered or platy texture. The layers can be split into thin sheets usually causing foliation in rocks.

◀ Hematite with reniform habit.

▼ Native copper is pure copper as found in nature. The isometric crystals of this sample, forming a dendritic pattern, are rare.

OTHER IDENTIFICATION PROPERTIES

Cleavage and fracture

As well as influencing the shape of a mineral crystal, the arrangement of atoms in the mineral structure determines the way the crystal breaks. The strength of the bonds that hold atoms together in a mineral's chemical structure varies. In the mica group minerals, for example, the bonds between the layers of atoms are much weaker than those within the layers. Micas therefore cleave readily into thin sheets. Cleavage is a repeatable

breakage related to the mineral's internal atomic structure, producing reasonably flat surfaces that reflect light consistently. Certain words, such as perfect, distinct and poor, are used to describe cleavage. Not all minerals exhibit cleavage.

Fracture, however, is not closely related to internal structure and usually produces irregular shapes. This property can be described as uneven, irregular or conchoidal (curved). A hackly fracture has sharp edges. Unlike cleavage, fracture cannot be repeated exactly.

Hardness

This important identification property requires a simple test to be carried out. The property is the resistance a mineral's surface shows to being scratched. Hardness is a product of atomic bonding and how closely atoms are packed in the mineral structure. A scale of hardness was set up in 1812 by Friedrich Mohs. For its 10 defining points, this scale has 10 relatively common minerals. However, the intervals between the points on the scale are not equal. In order to establish a mineral's hardness, a specimen is

▼ Cleavage is the tendency of a material to split apart along flat, smooth planes. These are planes of weak atomic bonding within the material. Cleavage is described as perfect, good, fair, poor, distinct or indistinct. Splitting along irregular flat planes is sometimes called 'parting', but this may not be apparent in all specimens of a given mineral. Many minerals have perfect cleavage, and in some cases in more than one direction. Octahedral cleavage in fluorite (calcium fluoride) allows material to be broken into perfect octahedral shapes. This can sometimes be misleading, since the octahedron is also a typical shape of fluorite crystals. Expertise is needed to distinguish cleavage surfaces from true crystal surfaces.

◀ ▼ Octahedral cleavage in fluorite.

◀ Massive quartz specimen with no obvious crystal form and uneven fracture.

tested against objects whose hardness is known, beginning with the softest and working up the scale until the mineral can be marked. Sometimes it is convenient to express hardness as a fraction between two of the whole numbers.

The Mohs hardness scale:
1. Talc
2. Gypsum
3. Calcite
4. Fluorite
5. Apatite
6. Orthoclase feldspar
7. Quartz
8. Topaz
9. Corundum
10. Diamond

▶ Smoky quartz crystal from Bradenberg, Namibia, showing a characterisitc hexagonal prismatic habit and conchoidal fracture.

Specific gravity

This is a useful identification property. The specific gravity of a mineral held in the hand can, with experience, be estimated. Quartz, a very common mineral, has a specific gravity of 2.65, while that of gold is 19.30. The specific gravity of many common minerals is around 3.00, but some, such as galena (7.58) and barite (4.50), are noticeably higher. Specific gravity depends on the packing and type of atoms in the mineral and is defined as the comparison between the weight of the mineral specimen and that of an equal volume of water. This test is relatively easy to carry out by displacement of water, using a eureka can.

Colour

Looking at the colour of a mineral in sunlight can help to identify it, as many minerals exhibit specific colours. However, it can also be very complex and can sometimes be misleading, as some minerals can show a great variety of colours: quartz, for example, can be colourless, white, purple, pink, yellowish and brown. There

▲ ▶ Galena from Herodsfoot Mine, Cornwall, England, UK and calcite (right). Though they are the same size their specific gravities are very different.

are a great number of common minerals that are white, colourless or greyish, and the colour of certain minerals changes when viewed under different light sources such as fluorescent or incandescent light. Colour is a product of the way incoming light is absorbed by the mineral. The parts of the spectrum not absorbed give the mineral its colour. Both chemical impurities and defects within the atomic structure of a mineral can influence the colouring, as can the actual elements of which the mineral is dominantly composed.

Streak

This is the colour of the mineral when observed in its powdered form. The easiest way of obtaining the powder of a mineral is by rubbing part of it across a streak plate (an unglazed white porcelain tile). A line, which is the mineral's streak, is left on the plate. However, some minerals are too hard, and the powder produced by this test is that of the plate itself. In this case, it is necessary to crush a small part of the mineral to be tested, or rub it against a much harder surface such as diamond sandpaper. Streak is a useful test, as it is consistent, even for a mineral that shows many colour varieties; all the various colour forms of quartz, for example, give a white streak. Many other common minerals, however, produce a white streak.

Lustre

The reflectiveness of a mineral's surface is its lustre. This can be of varying intensity and quality. Several terms are used to describe lustre: vitreous is a common lustre, with light reflected as from a glass surface; adamantine lustre is a brilliant reflection like that of diamond; metallic lustre gives a sheen like that of a metal. Terms including dull, earthy, resinous, waxy, silky and pearly are also used and are self-explanatory.

Transparency

The transparency of a mineral is usually considered during its identification. Minerals can be transparent, translucent or opaque.

Other properties

Some minerals will react with a variety of acids. Testing should be done carefully. The application of weak hydrochloric acid causes a definite reaction in certain minerals: calcite and other carbonate minerals effervesce, liberating carbon dioxide gas, while sulphides, including galena, produce foul-smelling hydrogen sulphide gas.

Fluorescence is a variable and often very colourful property, produced by placing a specimen under ultraviolet light. The colour of the fluorescence can be diagnostic. Both short-wave and long-wave ultraviolet light are used, because sometimes the colours observed will be different under these two sources of illumination. Testing should be done with appropriate training, as ultraviolet light can damage the eyes. Not all minerals exhibit fluorescence.

Native elements

These are minerals composed of chemical elements which exist alone and not combined with other chemical elements to form compounds. They are metallic elements such as gold and copper; semi-metals like antimony and arsenic; and non-metals such as sulphur and carbon.

Gold

Chemical composition: Au **Crystal system:** Cubic **Mineral habit:** Gold very rarely forms as cubic or octahedral crystals; more often it occurs as small nuggets, grains, flakes or in dendritic masses **Cleavage:** None **Fracture:** Hackly **Hardness:** 2½ to 3 **Specific gravity:** 19.30 **Colour:** Rich yellow **Streak:** Yellow-gold **Lustre:** Metallic; opaque

Often occurring in quartz veins, gold forms with other hydrothermal minerals. This very dense mineral is also found in alluvial sands in placer deposits. Gold accumulates in these deposits because, during weathering and erosion of the original quartz-bearing rocks, it is too dense to be washed away. For this reason, panning in streams and rivers is a long-standing method of obtaining the valuable metal, and grains and nuggets can be won in this way. It is thought that the world's oceans contain over 10 million tonnes of gold, but no economically viable method of obtaining this has yet been found. Gold can sometimes be confused with the metallic sulphide pyrite, often known as 'fool's gold', and chalcopyrite, but properties including hardness and specific gravity can overcome any identification problems. Gold is insoluble in most acids but can be dissolved in selenic acid and aqua regia, or in high-pressure water heated to 375°C (707°F). It doesn't tarnish on contact with air, as silver does. For thousands of years gold has been used in jewellery and as currency. Important areas where gold is mined include South Africa, Australia, the USA, Canada, Russia, Peru, Mexico, Chile and Brazil.

◄ Gold in quartz from West Springs mine, Transvaal, South Africa.

Silver

Chemical composition: Ag **Crystal system:** Cubic **Mineral habit:** Rarely as octahedral or cubic crystals; more often as wires, scales, sheets and dendritic or massive forms **Cleavage:** None **Fracture:** Hackly **Hardness:** 2½ to 3 **Specific gravity:** 10.50 **Colour:** Whitish-silver, tarnishing on exposure **Streak:** Silvery **Lustre:** Metallic; opaque

▶ Native silver from Kongsberg, Buskerud, Norway.

Forming in hydrothermal veins, silver commonly occurs with gold, metallic sulphides and other minerals containing silver. These include stephanite and acanthite. Silver may also form an alloy with gold, called electrum. Most silver is obtained from its various ores rather than as a native element. Unlike gold, silver can be dissolved in nitric acid. It tarnishes in the atmosphere. Peru, Mexico, Chile, Bolivia, the USA, Canada, Australia, Russia, Norway and the Czech Republic are all important sources of silver.

Platinum

Chemical composition: Pt **Crystal system:** Cubic **Mineral habit:** Very rare crystals are cubic; more commonly platinum occurs as scales, grains and nuggets **Cleavage:** None **Fracture:** Hackly **Hardness:** 4 to 4½ **Specific gravity:** 21.44 **Colour:** White or silver-grey **Streak:** White or greyish **Lustre:** Metallic; opaque

◀ Platinum nugget from Nijni-Tagilsk, Perm, Ural Mountains, Russia.

Though very rare, platinum forms in igneous rocks, especially those low in silica, of basic and ultrabasic composition, including dunite and the metamorphic rock serpentinite. Uncommonly it may be found as an alloy with iron and other elements. As with gold and other minerals of high specific gravity, eroded grains of platinum can be deposited in placers as flowing water loses its energy. The best known are those in the Ural Mountains of Russia, which are still worked. Other notable areas for platinum include South Africa (especially the Marensy Reef), Ontario in Canada, the USA, Colombia, Brazil, Peru, Australia and New Zealand. This metal is extremely ductile, more so than copper, silver or gold. It resists corrosion, but can oxidize at temperatures of over 500°C (932°F). Platinum is soluble only in aqua regia.

Copper

Chemical composition: Cu **Crystal system:**
Cubic **Mineral habit:** Rarely as cubic or octahedral
crystals; more often in the form of irregular
and dendritic masses, and wires **Cleavage:**
None **Fracture:** Hackly **Hardness:** 2½ to 3
Specific gravity: 8.94 **Colour:** Copper-red,
turning brown when tarnished **Streak:** Copper-
red **Lustre:** Metallic; opaque

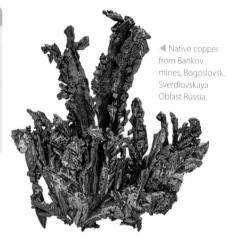

◀ Native copper
from Bankov
mines, Bogoslovsk,
Sverdlovskaya
Oblast Russia.

Copper is less common as a native element than its
ores, occurring in basaltic volcanic rocks. It is more
often found as a compound with other elements,
especially sulphur, forming sulphide minerals such
as chalcopyrite. Copper is a valuable mineral for
the manufacture of pipes and electrical wiring. In
large industrial quantities, it is mainly obtained from
sulphide ores. Copper is soluble in nitric acid and
is very malleable. The USA, Canada, Mexico, Bolivia,
the Mansa mine in Chile, Broken Hill in Australia, the
Rio Tinto mine in Palabora in South Africa, Namibia,
Botswana, Uganda, Zambia, Russia and Indonesia all
produce large quantities of copper.

Bismuth

Chemical composition: Bi **Crystal system:**
Trigonal **Mineral habit:** Usually occurring as
grains or massive aggregates and in lamellar,
foliated and dendritic forms, only rarely as
large crystals **Cleavage:** Perfect **Fracture:**
Uneven **Hardness:** 2 to 2½ **Specific gravity:**
9.70 to 9.83 **Colour:** Silvery white or pinkish,
with an iridescent blue tarnish **Streak:** Silver-
white **Lustre:** Metallic; opaque

◀ Bismuth from
Wolfram Camp,
Queensland,
Australia.

A mineral of hydrothermal veins, bismuth may
be found in deposits mined for silver, tin, nickel,
lead and cobalt. It can also occur in very coarse-
grained igneous pegmatites. Bismuth dissolves in
nitric acid. An opaque mineral, bismuth is readily
identified by its low hardness, malleability, high
specific gravity and colour. Bismuth has a high
value industrially, as it is a very poor conductor of
heat. It is usually obtained from mineral ores such
as bismuthinite and as a by-product from smelting
lead and copper ores. Man-made bismuth crystals
have characteristic 'hoppered' shapes and strongly
iridescent colours.

Arsenic

Chemical composition: As **Crystal system:** Trigonal **Mineral habit:** Usually as grains, botryoidal and stalactitic masses; rarely as crystals with rhombohedral form or as nodules **Cleavage:** Perfect **Fracture:** Uneven **Hardness:** 3½ **Specific gravity:** 5.72 to 5.73 **Colour:** Dull pale greyish, tarnishing to dark grey, almost black. **Streak:** Tin-white or greyish **Lustre:** Metallic; opaque

▲ Arsenic from Sacaramb, Romania.

This mineral forms in hydrothermal veins in association with sulphides and arsenides, especially those containing nickel, cobalt and silver. The element arsenic is a component of the colourful red mineral realgar and the related golden-coloured arsenic sulphide, orpiment. A well-known heating test produces fumes smelling of garlic. Arsenic is a poison and has been used in pesticides, though this has now largely stopped because of the problems caused to public health and the environment. However, large areas in parts of the USA are still contaminated.

Antimony

▼ Antimony from Arechuybo, Chihuahua, Mexico.

Chemical composition: Sb **Crystal system:** Trigonal **Mineral habit:** Generally found as massive aggregates, or in granular and lamellar forms; rare crystals have a pseudo-cubic, tabular, lath-like or acicular habit **Cleavage:** Perfect **Fracture:** Uneven **Hardness:** 3 to 3½ **Specific gravity:** 6.69 **Colour:** Silvery-white to grey **Streak:** Greyish **Lustre:** Brilliant metallic; opaque

Antimony occurs in hydrothermal mineral veins, usually associated with silver, arsenic, galena, sphalerite, pyrite and stibnite. This metallic mineral is used as an alloy in batteries and cable coatings. A moderately dense metal with low hardness, antimony is rare as a native element. As a chemical element it is found in many minerals, with stibnite, the main ore mineral, being the most common.

Mercury

Chemical composition: Hg **Crystal system:** Trigonal **Mineral habit:** Crystals only form at -38°C (-36.5°F), exhibiting a rhombohedral habit. At normal temperatures this metal is a liquid, occurring as small globules on the surface of rocks **Cleavage:** None (at room temperature) **Fracture:** None (at room temperature) **Hardness:** Not obtainable (at room temperature) **Specific gravity:** 14.38 **Colour:** Silvery-white **Streak:** None (at room temperature) **Lustre:** Brilliant metallic; opaque

Mercury is often associated with its red-coloured sulphide mineral, cinnabar. These occur around hot springs and volcanic vents. Native mercury

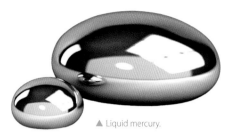

▲ Liquid mercury.

is often formed by the alteration of cinnabar deposits. It will dissolve in nitric acid. This mineral was formerly used in thermometers, but this use has been discontinued, as it is highly poisonous. Mercury is often known as 'quicksilver', a reference to it being a silver-coloured liquid.

Diamond

Chemical composition: C **Crystal system:** Cubic **Mineral habit:** Diamond forms as small crystals with octahedral, tetrahedral or cubic habit. A variety with a micro-crystalline habit is called carbonado and bort is the name given to radiating masses **Cleavage:** Perfect **Fracture:** Conchoidal **Hardness:** 10 **Specific gravity:** 3.51 **Colour:** Diamond can exhibit a variety of colours, ranging from colourless and white, to grey, yellowish, pink, reddish, orange, blue, green and black **Streak:** White **Lustre:** Adamantine or greasy; transparent to opaque

▶ Diamond from Oven's Diggings, Victoria, Australia.

Diamond forms naturally in a rock called kimberlite, named after the famous diamond locality in South Africa. This ultrabasic rock is found in deep, pipe-shaped, intrusive structures. Some diamonds also occur in lamprolite, another volcanic rock. The majority of diamonds were formed at depths of up to 250 km (155 miles) below the surface, well under the Earth's crust, in the region called the mantle. Their formation from fluids under great pressure may have taken place during Pre-Cambrian times, over 1,000 million years ago. Diamonds have been brought nearer to the surface more recently, possibly a few hundred million years ago. Because of its great hardness (defining point ten on Mohs hardness scale), diamond can resist erosion and accumulate in alluvial placer deposits and occur in beach sands.

Graphite

Chemical composition: C **Crystal system:** Hexagonal **Mineral habit:** Commonly forms in massive crystalline aggregates and earthy or granular habits, sometimes with a foliated texture. It can also occur as flattened crystal plates with a hexagonal outline. **Cleavage:** Perfect **Fracture:** Uneven **Hardness:** 1 to 2 **Specific gravity:** 2.09 to 2.23 **Colour:** Dark grey to black. **Streak:** Dark grey to black **Lustre:** Dull metallic; opaque

▲ Graphite from Ragedara, Kurunegala District, Sri Lanka.

Graphite is chemically the same as diamond, but the two minerals are surprisingly different in appearance and properties. For example, diamond is the measure of point 10 on the Mohs hardness scale (the hardest), while graphite is easily scratched with a finger nail, having a hardness of only 1 to 2; when it is drawn across a sheet of paper, a dark grey mark is created. Diamond is a brilliantly lustrous crystalline mineral, while graphite is dark grey to black and

often found in foliated masses. Graphite generally forms in low- and medium-grade metamorphic rocks, especially slate and schist, and because of its structure with parallel layers of carbon atoms, it has an industrial use in lubricants.

Sulphur

▼ Sulphur from Cesena, Emilia-Romagna, Italy.

Chemical composition: S **Crystal system:** Orthorhombic **Mineral habit:** Often forms as tabular crystals in small masses and more rarely as prismatic crystals; also as coatings on rock surfaces and in stalactitic, massive and encrusting habits **Cleavage:** Imperfect **Fracture:** Uneven or conchoidal **Hardness:** 1½ to 2½ **Specific gravity:** 2.07 **Colour:** Bright yellow to brownish yellow **Streak:** White **Lustre:** Resinous to greasy; transparent to translucent.

Sulphur usually occurs around volcanic vents and hot springs, where it is formed by gases and fluids issuing from depth. The mineral is also created by the action of bacteria on sulphate group minerals, including gypsum, in salt domes. Sulphur readily combines with metallic elements to produce a great number of sulphide and sulphate minerals,

such as pyrite, galena, sphalerite and barite. Sulphur has many industrial applications, including the manufacture of sulphuric acid, agricultural pesticides and the vulcanization of rubber.

Sulphides, sulphosalts, arsenides and tellurides

Sulphide minerals, a very important group of metallic ores, are chemical compounds of sulphur and metals. With similar properties to the sulphides, sulphosalts are compounds of sulphur with metallic and semi-metallic elements, such as arsenic. If, as is the case with sylvanite and skutterudite, tellurium or arsenic is substituted for sulphur, a telluride or arsenide is produced.

Sulphides

Galena

Chemical composition: PbS **Crystal system:** Cubic **Mineral habit:** Cubic and octahedral crystals are common; also occurs in massive and granular habits **Cleavage:** Perfect **Fracture:** Subconchoidal **Hardness:** 2½ **Specific gravity:** 7.58 **Colour:** Lead-grey **Streak:** Grey **Lustre:** Metallic; opaque

Galena is a common mineral that has been mined for its lead content for thousands of years. It is generally found in mineral veins and is formed from hydrothermal fluids, along with other common sulphides such as sphalerite and pyrite and also with quartz, fluorite and calcite. Galena may contain silver; as much as 1.2 kg (2½ lb) of silver per tonne of galena has been recorded. When placed in hydrochloric acid, hydrogen sulphide gas is produced, giving the characteristic smell of rotten eggs. Galena is found worldwide.

Clausthalite forms a series with galena. It is lead selenide (PbSe) and resembles galena in terms of its greyish colour and metallic lustre. It has a slightly higher specific gravity, 8.08 to 8.22, and a hardness of 2½ to 3. Clausthalite is classified in the cubic system, but it usually forms in a massive, rather than crystalline, habit. It occurs in hydrothermal veins, often associated with other selenides. This mineral was first described from the Harz Mountains in Germany.

◀ Galena from Galena, Cherokee County, Kansas, USA.

Cinnabar

Chemical composition: HgS **Crystal system:** Trigonal **Mineral habit:** Often as crystals on rock surfaces, with prismatic, rhombohedral or tabular habits; more usually in massive aggregates and as small grains **Cleavage:** Perfect **Fracture:** Conchoidal or uneven **Hardness:** 2 to 2½ **Specific gravity:** 8.08 **Colour:** Red-brown to scarlet **Streak:** Red **Lustre:** Vitreous, adamantine, submetallic or dull; transparent to opaque

▼ Cinnabar from eastern Slovakia.

Cinnabar can occur around volcanic vents and hot springs in association with native mercury and other minerals, including stibnite, quartz, marcasite and opal. It is also found in mineral veins and sedimentary rocks where volcanic activity has occurred. Cinnabar is the most important ore of mercury, but because of the high toxicity of this element, some of its traditional uses (as a red pigment, for example) are now discontinued and other materials have been substituted. There are many cinnabar deposits in North and South America and the Almaden region of Spain has been mined for cinnabar for 2,500 years.

Greenockite

Chemical composition: CdS **Crystal system:** Hexagonal **Mineral habit:** Usually forms on other minerals as a coating with an earthy appearance; can also be found as small pyramidal, tabular or prismatic crystals, often with striated faces **Cleavage:** Distinct **Fracture:** Conchoidal **Hardness:** 3 to 3½ **Specific gravity:** 4.82 **Colour:** Orange, yellow-orange, yellow or reddish **Streak:** Orange-yellow to brick-red **Lustre:** Adamantine or resinous; transparent to translucent.

▼ Greenockite from Bishopton Tunnel, Bishopton, Renfrewshire, Scotland, UK.

Greenockite forms in ore deposits in association with sphalerite and is often a direct alteration product of this mineral when it contains cadmium. In some instances it occurs as crystals in fissures and cavities in basic igneous rocks, associated with prehnite, zeolites and calcite.

Although quite rare, greenockite is the most common cadmium-rich mineral. This mineral was first discovered in Scotland.

Acanthite

Chemical composition: Ag_2S **Crystal system:**
Monoclinic **Mineral habit:** Prismatic crystals
Cleavage: None **Fracture:** Uneven
Hardness: 2 to 2½ **Specific gravity:** 7.22
Colour: Dark grey to black **Streak:** Black
Lustre: Metallic; opaque

Acanthite forms in hydrothermal veins with native
silver and other sulphide minerals, especially
galena, and is often associated with cerussite,
pyrargyrite and other hydrothermal vein minerals.
It is the main ore of silver, with important mines in
Mexico, Bolivia and Honduras.

▲ Acanthite from
Wheal Newton,
Harrowbarrow, near
Calstock, Cornwall,
England, UK.

Cobaltite

Chemical composition: CoAsS **Crystal
system:** Orthorhombic **Mineral habit:**
Crystals striated octahedral, or pseudocubic;
also granular or massive **Cleavage:** Perfect
Fracture: Uneven **Hardness:** 5½
Specific gravity: 6.33 **Colour:** Dark grey,
bluish or white **Streak:** Dark grey **Lustre:**
Metallic; opaque

▼ Cobaltite from
Tunaberg, Sweden.

Cobaltite occurs in hydrothermal veins and
in rocks altered by thermal metamorphism,
where it is often accompanied by magnetite,
sphalerite and titanite. It is soluble in nitric acid.
Cobaltite is part of a series with gersdorffite
(NiAsS), a mineral in the cubic system, with a
hardness of 5½ and a specific gravity of 5.90.
The colour is tin-white to grey, often tarnishing
to dark grey, which is the colour of the streak.
This is an opaque mineral, with a metallic
lustre. Like cobaltite, gersdorffite is found in
hydrothermal veins, where both minerals are
associated with chalcopyrite, calcite, dolomite
and quartz.

Sphalerite

Chemical composition: ZnS **Crystal system:** Cubic **Mineral habit:** Crystals are common and show tetrahedral and dodecahedral habits; also massive, botryoidal and granular **Cleavage:** Perfect **Fracture:** Uneven or conchoidal **Hardness:** 3½ to 4 **Specific gravity:** 3.90 to 4.10 **Colour:** Dark brown to black, reddish, yellowish, greenish or grey **Streak:** Pale brown **Lustre:** Vitreous or resinous; translucent to transparent

▼ Sphalerite from Rodna Veche, Transylvania, Romania.

Sphalerite is a common mineral of hydrothermal veins and in many areas it has been mined for its zinc content for hundreds of years. It usually occurs with quartz, galena, calcite, barite, chalcopyrite and fluorite. Wurtzite, an uncommon mineral, is dimorphous with sphalerite and forms pyramidal, prismatic or tabular crystals in the hexagonal crystal system. It is otherwise similar in its properties to sphalerite.

Stibnite

Chemical composition: Sb_2S_3 **Crystal system:** Orthorhombic **Mineral habit:** Crystals prismatic, often striated; also granular and bladed habits **Cleavage:** Perfect **Fracture:** Uneven or subconchoidal **Hardness:** 2 **Specific gravity:** 4.63 to 4.66 **Colour:** Grey **Streak:** Grey **Lustre:** Metallic; opaque

This mineral forms in hydrothermal veins and around hot springs. In veins, it is often associated with pyrite, sphalerite, galena, barite and calcite. It has also been found in the igneous rock, granite, and the high-grade regionally metamorphosed rock, gneiss. It is not uncommon for the prismatic crystals to be twisted due to the mineral's malleable nature. Stibnite can be dissolved in hydrochloric acid. It is the main ore of antimony.

◄ Stibnite from Hunan, China.

Bornite

Chemical composition: Cu_5FeS_4 **Crystal system:** Orthorhombic **Mineral habit:** Bornite may form as octahedral, cubic and dodecahedral crystals, which frequently have curved faces; it commonly occurs as massive or granular forms **Cleavage:** Poor **Fracture:** Uneven to conchoidal **Hardness:** 3 **Specific gravity:** 5.08 **Colour:** Coppery-red to brown. It tarnishes easily to red, blue and purple, and is often called 'peacock ore' **Streak:** Greyish to black **Lustre:** Metallic; opaque

▶ Bornite from Cumbria, England, UK.

Bornite forms in hydrothermal veins, often associated with quartz and sulphide minerals, especially galena. It also occurs in some contact metamorphic rocks and in coarse granitic pegmatites. In many copper deposits, bornite and chalcopyrite may be replaced by chalcocite. Bornite is an important ore of copper, often containing over 60% of the metal. This mineral is soluble in nitric acid.

Chalcopyrite

Chemical composition: $CuFeS_2$ **Crystal system:** Tetragonal **Mineral habit:** Commonly massive, but also as pseudotetrahedral crystals, often with striations on their faces, or in reniform shapes **Cleavage:** Poor **Fracture:** Uneven **Hardness:** 3½ to 4 **Specific gravity:** 4.35 **Colour:** Brassy yellow, tarnishing iridescent **Streak:** Greenish-black **Lustre:** Metallic; opaque

granite bodies. This mineral is a very important ore of copper. After it has been heated, it may become magnetic. Like bornite, chalcopyrite is soluble in nitric acid, but has a lower specific gravity and is slightly harder.

A common and widespread hydrothermal sulphide, chalcopyrite forms in mineral veins, in combination with other minerals such as galena, pyrite, sphalerite and quartz. It also occurs in some nodular masses of pyrite and in concentrated copper deposits in

▶ Chalcopyrite on quartz crystals from Cornwall, England, UK.

Chalcocite

Chemical composition: Cu$_2$S **Crystal system:** Monoclinic **Mineral habit:** Usually occurs in a massive habit, but occasionally forms short, tabular or prismatic crystals. When twinned, these may have a pseudohexagonal appearance **Cleavage:** Indistinct **Fracture:** Conchoidal **Hardness:** 2½ to 3 **Specific gravity:** 5.50 to 5.80 **Colour:** Grey to almost black **Streak:** Dark grey **Lustre:** Metallic; opaque

▼ Chalcocite from St. Ives Consols, Cornwall, England, UK.

Chalcocite occurs in hydrothermal mineral veins with a variety of sulphides and other minerals, including bornite, chalcopyrite, galena, malachite, azurite, covellite and quartz. It forms as a secondary mineral by the alteration and oxidation of primary copper minerals and is important in large-scale copper ore deposits. As with a number of other copper sulphide minerals, chalcocite dissolves in nitric acid.

Covellite

Chemical composition: CuS **Crystal system:** Hexagonal **Mineral habit:** Commonly as massive, foliated aggregates, but also occurs as thin, tabular plates with a hexagonal outline **Cleavage:** Perfect **Fracture:** Uneven **Hardness:** 1½ to 2 **Specific gravity:** 4.68 **Colour:** Dark blue, often with a purple iridescence **Streak:** Dark grey to black **Lustre:** Dull or submetallic; opaque

▼ Covellite from Butte, Montana, USA.

Covellite is usually found in mineral veins with other copper minerals, especially in zones where alteration of primary minerals has taken place. It often occurs as thin coatings on other minerals. In copper containing veins, it forms at relatively shallow depth within the zone of secondary enrichment. Covellite is commonly found with bornite, pyrite, chalcocite and chalcopyrite. This mineral may be formed when metamorphism of rocks containing copper minerals takes place. It is soluble in hydrochloric acid.

Orpiment

Chemical composition: As_2S_3 **Crystal system:** Monoclinic **Mineral habit:** As massive aggregates or in small foliated masses; columnar and botryoidal shapes are also found; crystals have a prismatic habit but are rare **Cleavage:** Perfect **Fracture:** Uneven **Hardness:** 1½ to 2 **Specific gravity:** 3.49 **Colour:** Golden yellow or brownish yellow **Streak:** Pale yellow **Lustre:** Pearly or resinous; transparent to translucent

▼ Orpiment.

Orpiment can be found around volcanic vents and hot springs as well as in hydrothermal veins. It may form by the alteration of realgar, another brightly coloured sulphide of arsenic. It often occurs with a variety of minerals including realgar, native arsenic, calcite, barite, stibnite and quartz. Orpiment is readily dissolved in nitric acid. If heated, it gives off a strong smell of garlic; this smell is commonly associated with minerals containing arsenic.

Realgar

Chemical composition: As_4S_4 **Crystal system:** Monoclinic **Mineral habit:** Short, prismatic crystals with striated faces; also in massive and granular habits, as foliated masses and powdery coatings **Cleavage:** Distinct **Fracture:** Conchoidal **Hardness:** 1½ to 2 **Specific gravity:** 3.56 **Colour:** Orange-red or red **Streak:** Orange-red **Lustre:** Greasy or resinous; transparent to translucent

This mineral occurs with orpiment, stibnite and a variety of other minerals, in hydrothermal veins. It can also crystallize from fluids issuing from hot springs, as in the geyser deposits in Yellowstone National Park, USA. Realgar gives off the smell of garlic when heated, characteristic of arsenic-containing minerals. and it dissolves in nitric acid. If exposed to light for any length of time, realgar crystals will break down to become a yellow-orange powder.

▶ Realgar.

Molybdenite

Chemical composition: MoS_2 **Crystal system:** Hexagonal **Mineral habit:** Crystals barrel-shaped or tabular; also as thin scales and foliated masses, often with a hexagonal outline or as grains **Cleavage:** Perfect **Fracture:** Uneven **Hardness:** 1 to 1½ **Specific gravity:** 4.62 to 4.73 **Colour:** Silvery grey **Streak:** Grey **Lustre:** Metallic; opaque

▼ Molybdenite from Slangsvoll, Råde, Østfold, Norway.

Molybdenite is found as an accessory mineral in some granites and pegmatites and also occurs in hydrothermal veins and in rocks altered by contact metamorphism. When handled, specimens feel greasy and leave small silvery flakes on the hands. It has a similar general appearance to graphite, but is more metallic, with a higher specific gravity.

This is a mineral of some economic value, being the main ore of molybdenum, which has many uses including as a lubricant. It is mined in USA, Canada, Chile and Russia.

Bismuthinite

Chemical composition: Bi_2S_3 **Crystal system:** Orthorhombic **Mineral habit:** Crystals acicular or prismatic; rarely in massive, foliated or fibrous habits **Cleavage:** Perfect **Fracture:** Uneven **Hardness:** 2 **Specific gravity:** 6.78 **Colour:** Silvery white to lead-grey **Streak:** Lead-grey **Lustre:** Metallic; opaque

Occurring with a variety of other sulphide minerals, including pyrite, galena, arsenopyrite and chalcopyrite, bismuthinite is found in hydrothermal veins and in granite pegmatites, where it may be associated with copper minerals. It also forms in some gold-bearing veins and volcanic rocks and is an important ore of bismuth. Bismuthinite dissolves in nitric acid, and small flakes of sulphur may appear in the liquid. The main mining areas for this mineral are Bolivia, Peru, Mexico, Canada and Japan.

▶ Bismuthinite from Fowey Consols Mine, Cornwall, England, UK.

Marcasite

Chemical composition: FeS₂ **Crystal system:**
Orthorhombic **Mineral habit:** Tabular or
pyramidal crystals, commonly occurring in
spear-shaped aggregates when twinned; also
massive or reniform; flattened masses with a
radiating structure are known as marcasite 'suns'.
Nodules of marcasite have an internal radiating
structure **Cleavage:** Distinct **Fracture:**
Uneven **Hardness:** 6 to 6½ **Specific gravity:** 4.89
Colour: Pale brass-yellow **Streak:** Greenish-
black **Lustre:** Metallic; opaque

Marcasite has the same chemical formula as
pyrite but differs in its crystal system. Both
minerals have similar hardness, but marcasite is
paler coloured. This mineral commonly forms in
sedimentary rocks, including clay and limestone,
where it can be produced by the action of
acidic water on the rock. It also occurs in some
hydrothermal veins, often with pyrite and other
sulphides. On exposure to the air, it decomposes
more rapidly than pyrite. Some well-known
occurrences are in the Cretaceous rocks of
southern England and France.

▼ Marcasite from
near Dover, Kent,
England, UK showing
the twinned spear
shapes.

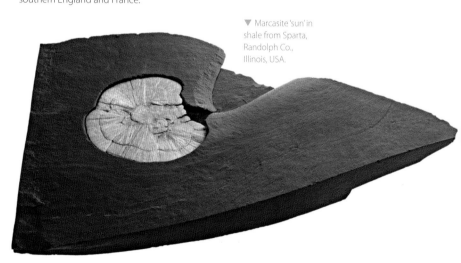

▼ Marcasite 'sun' in
shale from Sparta,
Randolph Co.,
Illinois, USA.

Pyrite

Chemical composition: FeS_2 **Crystal system:** Cubic **Mineral habit:** Cubic, octahedral and pyritohedral crystals, the faces of which are often striated; also granular, massive, reniform, nodular **Cleavage:** Indistinct **Fracture:** Uneven or conchoidal **Hardness:** 6 to 6½ **Specific gravity:** 4.80 to 5.00 **Colour:** Brass yellow **Streak:** Greenish-black **Lustre:** Metallic; opaque

A very common, widespread sulphide mineral, pyrite forms in many geological environments. It occurs in hydrothermal veins and as an accessory mineral in igneous and regionally metamorphosed rocks, especially slates. It is also found in sedimentary rocks, often forming in rounded nodules, and can replace organic material during fossilization. When struck with a geological hammer, sparks may

▼ Pyrite from Tamajun, Soria, Castile and Leon, Spain.

be produced. The name 'fool's gold' has been applied to this mineral, but it is far harder than gold and has a much lower specific gravity. It can be confused with chalcopyrite, but pyrite is harder and not as rich a colour. Pyrite can decompose on exposure to the air.

Pyrrhotite

Chemical composition: FeS **Crystal system:** Monoclinic **Mineral habit:** Platy or tabular crystals, sometimes twinned; usually massive, granular **Cleavage:** None **Fracture:** Subconchoidal to uneven **Hardness:** 3½ to 4½ **Specific gravity:** 4.58 to 4.65 **Colour:** Bronze-yellow to bronze-red, often tarnishing to brown with iridescent colouring **Streak:** Dark grey or black **Lustre:** Metallic; opaque

This mineral has a very similar chemical composition to pyrite, but often occurs in magmatic intrusions, particularly those of basic and ultrabasic composition, usually only in small quantities, with pyrite, magnetite, marcasite, galena and sphalerite. In intrusions which have a layered structure, it may be found with chalcopyrite and pentlandite. Pyrrhotite also

▶ Pyrrhotite from Santa Eulalia, Chihuahua, Mexico.

forms in rocks altered by contact metamorphism. Because of its magnetism, and its propensity to replace pyrite, this mineral has been called 'magnetic pyrite'.

Stannite

Chemical composition: Cu_2FeSnS_4 **Crystal system:** Tetragonal **Mineral habit:** Pseudo-tetragonal crystals, often striated and twinned; also granular, massive **Cleavage:** Indistinct **Fracture:** Uneven **Hardness:** 4 **Specific gravity:** 4.30 to 4.50 **Colour:** Steel-grey to black; it may have a bluish tarnish **Streak:** Black **Lustre:** Metallic; opaque

▼ Stannite from Wheal Agar, Carn Brea, Cornwall, England, UK.

An ore of tin, stannite occurs in hydrothermal veins with cassiterite, pyrite, arsenopyrite, sphalerite, chalcopyrite, wolframite and tetrahedrite. It is known from many of the world's tin-producing regions, especially Cornwall, UK, where it was first discovered, and is associated with cassiterite, another important tin ore.

Arsenopyrite

Chemical composition: FeAsS **Crystal system:** Monoclinic **Mineral habit:** Usually as twinned prismatic crystals, often with striated faces; also in columnar, granular and massive habits **Cleavage:** Distinct **Fracture:** Uneven **Hardness:** 5½ to 6 **Specific gravity:** 6.07 **Colour:** Silvery white, sometimes tarnishing iridescent, grey or yellowish **Streak:** Black **Lustre:** Metallic; opaque

This is a mineral of hydrothermal veins, where it is associated with many minerals including gold, silver, pyrite, siderite, chalcopyrite, quartz and calcite. Asenopyrite also occurs in basaltic igneous rocks and in metamorphic rocks. It is soluble in nitric acid and gives off sparks and a smell of garlic when struck with a geological hammer. Arsenopyrite is an important ore of arsenic.

▼ Arsenopyrite from Freiberg, Saxony, Germany.

Sulphosalts

Enargite

Chemical composition: Cu_3AsS_4 **Crystal system:** Orthorhombic **Mineral habit:** As well-formed tabular or prismatic crystals, often with striated faces; also granular and massive **Cleavage:** Perfect **Fracture:** Uneven **Hardness:** 3 **Specific gravity:** 4.45 **Colour:** Dark grey to black **Streak:** Grey to black **Lustre:** Metallic; opaque

This mineral occurs in hydrothermal mineral veins with quartz and sulphides such as galena, pyrite, sphalerite, chalcopyrite and bornite. When heated, a smell of garlic is produced. Enargite melts in a flame and is soluble in nitric acid.

▶ Enargite from Leonard Mine, Butte, Montana, USA.

Jamesonite

Chemical composition: $Pb_4FeSb_6S_{14}$ **Crystal system:** Monoclinic **Mineral habit:** Usually as fibrous and acicular crystals, striated along their length; also massive and as fibrous or columnar aggregates **Cleavage:** Good **Fracture:** Uneven to conchoidal **Hardness:** 2½ **Specific gravity:** 5.63 **Colour:** Dark grey to black, sometimes tarnishing iridescent **Streak:** Dark grey to black **Lustre:** Metallic; opaque

▼ Jamesonite from St. Minver, Cornwall, England, UK.

Jamesonite is unusual for a sulphosalt mineral in that its brittle fibrous or acicular crystals are sometimes so dense and slender that they form felt-like mats on rock surfaces. It occurs in hydrothermal veins, associated with other sulphosalts, sulphides, quartz and calcite. Jamesonite was first identified in Cornwall, UK.

Pyrargyrite

Chemical composition: Ag_3SbS_3 **Crystal system:** Trigonal **Mineral habit:** Prismatic and scalenohedal crystals, often twinned; also as disseminated grains, compact and massive **Cleavage:** Distinct and indistinct, depending on direction **Fracture:** Conchoidal to uneven **Hardness:** 2½ **Specific gravity:** 5.85 **Colour:** Dark red **Streak:** Dark red **Lustre:** Adamantine or submetallic; translucent

◀ Pyrargyrite from Andreasberg, Germany.

Pyrargyrite is closely related to proustite, having similar physical characteristics. However, pyrargyrite contains antimony, whereas proustite has arsenic in its chemical formula. Pyrargyrite forms in hydrothermal mineral veins with other sulphosalts, galena, pyrite, silver, calcite, dolomite and quartz. Pyrargyrite is one of the very few silver-bearing minerals that exhibit translucency and has been called 'ruby silver', because of its characteristic colouring. It is soluble in nitric acid. Excellent crystals have been found in the Harz Mountains in Germany, as well as in Mexico. Pyrargyrite also occurs in the USA and Canada, Chile, Bolivia, Peru, Spain, the Czech Republic and Slovakia.

Proustite

Chemical composition: Ag_3AsS_3 **Crystal system:** Trigonal **Mineral habit:** Prismatic, rhombohedral and scalenohedal crystals, often twinned; can be massive, compact, disseminated **Cleavage:** Distinct **Fracture:** Conchoidal to uneven **Hardness:** 2 to 2½ **Specific gravity:** 5.55 to 5.64 **Colour:** Scarlet, tarnishing to black **Streak:** Scarlet **Lustre:** Adamantine to submetallic; transparent to translucent

Proustite shares many physical characteristics with pyrargyrite but is a brighter red colour and can be transparent. It also has a slightly lower specific gravity. Proustite forms in hydrothermal veins, where it occurs with other sulphosalts, silver, galena, quartz, calcite, dolomite and pyrite. As with pyrargyrite, proustite dissolves in nitric acid and is easily melted.

▼ Proustite from Freiberg, Saxony, Germany.

Bournonite

Chemical composition: $CuPbSbS_3$ **Crystal system:** Orthorhombic **Mineral habit:** Prismatic and tabular, striated crystals, often twinned; also as compact, granular and massive forms **Cleavage:** Imperfect **Fracture:** Subconchoidal to uneven **Hardness:** 2½ to 3 **Specific gravity:** 5.83 **Colour:** Steel-grey to black **Streak:** Grey or black **Lustre:** Metallic; opaque

This mineral forms in hydrothermal veins in association with galena, sphalerite, siderite, chalcopyrite, stibnite, tetrahedrite and quartz. The twinned crystals can occur in cruciform (cross-shaped) groups, or in the rough shape of a cog wheel; the term 'cog-wheel ore' has previously been used for bournonite. This mineral dissolves in nitric acid, leaving a greenish tint indicative of its copper content. It melts very easily in a flame.

▲ Bournonite from Herodsfoot Mine, Cornwall, England, UK showing the cog-wheel crystal form.

Tetrahedrite

▼ Tetrahedrite from Pranal Mine, Pontgibaud, Puy-de-Dome, France.

Chemical composition: $Cu_6Cu_4(Fe,Zn)_2Sb_4S_{13}$ **Crystal system:** Cubic **Mineral habit:** Tetrahedral crystals, often twinned; also massive, compact and granular habits **Cleavage:** None **Fracture:** Uneven to subconchoidal **Hardness:** 3 to 4½ **Specific gravity:** 4.60 to 5.10 **Colour:** Grey to black **Streak:** Black to brown, or reddish **Lustre:** Metallic; opaque

Tetrahedrite forms a series with tennantite. It contains antimony, whereas tennantite contains arsenic. Tetrahedrite forms in hydrothermal veins with typical vein minerals such as sphalerite, galena, chalcopyrite, fluorite, barite, calcite and quartz. It also occurs in pegmatites. Tetrahedrite dissolves in nitric acid.

Tennantite

Chemical composition: $Cu_6Cu_4(Fe,Zn)_2As_4S_{13}$
Crystal system: Cubic **Mineral habit:**
Tetrahedral crystals, commonly twinned; also
massive, granular and compact **Cleavage:** None
Fracture: Uneven to subconchoidal
Hardness: 3 to 4½ **Specific gravity:** 4.59 to
4.75 **Colour:** Grey to black **Streak:** Black to
brown or dark red **Lustre:** Metallic; opaque

Being in a series with tetrahedrite, tennantite
shares many properties, and it is difficult to
distinguish between the two minerals visually.
Tennantite occurs in hydrothermal mineral veins
with pyrite and other sulphides as well as siderite,
quartz, barite, fluorite, calcite and dolomite. It
melts easily and is soluble in nitric acid.

▼ Tennantite
from Peru.

Arsenides

Nickeline

Chemical composition: NiAs **Crystal system:**
Hexagonal **Mineral habit:** Commonly in
reniform, massive, disseminated and columnar
habits; rare crystals are pyramidal and small
Cleavage: None **Fracture:** Uneven **Hardness:**
5 to 5½ **Specific gravity:** 7.78 **Colour:** Copper-
red, tarnishing greyish to blackish **Streak:**
Brownish-black **Lustre:** Metallic; opaque

Nickeline occurs in hydrothermal veins, often
with other nickel-containing minerals, and
those with silver and cobalt in their chemical
structure. It also forms in the basic igneous
rock, norite, a type of gabbro. Ore deposits
formed by the alteration of ultrabasic rocks
frequently contain nickeline along with

chalcopyrite, cobaltite, pyrrhotite, arsenopyrite
and barite. It is soluble in nitric acid. When
heated, a smell of garlic is produced due to its
arsenic content. An alternative name for this
mineral is niccolite.

▼ Nickeline from
Linares, Jaen,
Andalucia, Spain.

Skutterudite

Chemical composition: $CoAs_3$ **Crystal system:** Cubic **Mineral habit:** Commonly occurs in granular and massive habits; rare crystals may be cubic or octahedral **Cleavage:** Distinct **Fracture:** Uneven **Hardness:** 5½ to 6 **Specific gravity:** 6.10 to 6.90 **Colour:** Tin-white **Streak:** Black **Lustre:** Metallic; opaque

Skutterudite forms in hydrothermal mineral veins with minerals containing nickel and cobalt. Associated minerals include sulphosalts, arsenopyrite, nickeline, cobaltite, barite, calcite and siderite. Native silver may also be found with skutterudite. When tarnished, skutterudite can become grey or iridescent. A smell of garlic is produced when it is heated.

▶ Skutterudite from Richelsdorf, Hesse, Germany.

Sperrylite

Chemical composition: $PtAs_2$ **Crystal system:** Cubic **Mineral habit:** Crystals usually cubic or octahedral; also massive **Cleavage:** Indistinct **Fracture:** Conchoidal **Hardness:** 6 to 7 **Specific gravity:** 10.46 to 10.60 **Colour:** Pale tin-white **Streak:** Black **Lustre:** Metallic; opaque

This mineral is an arsenide within the pyrite group of minerals. It has a similar crystal structure to pyrite. Sperrylite occurs in rocks that have been subjected to contact metamorphism and, because of its high specific gravity, hardness and resistance to weathering, it is often found in river gravels and sands. It sometimes occurs in association with covellite. Sperrylite is the most widespread platinum-containing mineral and is found in the USA, Canada, Finland, Siberia, South Africa and Australia.

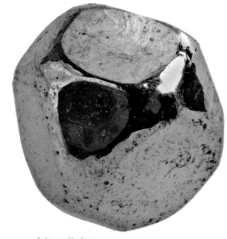

▲ Sperrylite from Tweefontein Farm, Waterberg district, Limpopo, South Africa.

Tellurides

Sylvanite

Chemical composition: $AuAgTe_4$ **Crystal system:** Monoclinic **Mineral habit:** Often as granular, crystalline and cleaved masses; crystals are tabular or prismatic and frequently twinned **Cleavage:** Perfect **Fracture:** Uneven **Hardness:** 1½ to 2 **Specific gravity:** 8.16 **Colour:** Silvery white, less often yellowish **Streak:** Silvery white **Lustre:** Metallic; opaque

This mineral contains both gold and silver in its chemical formula, and so has a high specific gravity. It forms in hydrothermal veins with gold, quartz, fluorite, pyrite and other sulphides, carbonates such as rhodochrosite and calcite, and other tellurides. It is soluble in nitric acid and will tarnish in strong light. If a specimen is heated in sulphuric acid, the liquid becomes red-coloured. .

▶ Sylvanite from Cripple Creek, Teller County, Colorado, USA.

Halides

The minerals in this group form when metallic elements combine with halogen elements such as fluorine, bromine and chlorine.

Halite

Chemical composition: NaCl **Crystal system:** Cubic **Mineral habit:** Crystals cubic, often with stepped concave faces (hopper crystals), rarely octahedral; also massive, compact and granular **Cleavage:** Perfect **Fracture:** Conchoidal or uneven **Hardness:** 2 **Specific gravity:** 2.17 **Colour:** Colourless, white, reddish, orange, yellow, blue, purple **Streak:** White **Lustre:** Vitreous, resinous; transparent to translucent

Halite occurs in evaporite deposits where saline water, such as a marine lagoon or inland salt lake, dries out. The various chemical salts held in solution in the water are precipitated in a sequence, the least soluble first and the most soluble, including halite, last. The evaporite minerals found with halite include, gypsum and sylvite. These are usually interbedded with clay and other sedimentary rocks. The deposits in which halite forms can be as little as a few centimetres thick or as much as hundreds of metres. Halite also occurs as a sublimation product around volcanic vents. It has a distinct salty taste and is the salt that we use in cooking and preserving. It can be dissolved in cold water.

▼ Halite from Wieliczka, Poland.

Sylvite

Chemical composition: KCl **Crystal system:** Cubic **Mineral habit:** Crystals cubic, rarely octahedral; also massive, encrusting and granular **Cleavage:** Perfect **Fracture:** Uneven **Hardness:** 2 **Specific gravity:** 1.99 **Colour:** Colourless, white, grey, yellow, bluish or purple; reddish with iron mineral inclusions **Streak:** White **Lustre:** Vitreous; transparent

▶ Sylvite.

Sylvite occurs in evaporite deposits together with halite, gypsum, anhydrite, carnallite and polyhalite, and as a volcanic sublimation product. It has a bitter taste and is soluble in cold water. It melts easily, colouring the flame purplish-red. As an important component in potash, sylvite has various industrial applications, including its use in agricultural fertilizers. Major deposits occur in Saskatchewan, Canada.

Chlorargyrite

Chemical composition: AgCl **Crystal system:** Cubic **Mineral habit:** Rare as cubic crystals; massive, flaky, as waxy coatings and crusts on rock surfaces **Cleavage:** None **Fracture:** Uneven or subconchoidal **Hardness:** 2½ **Specific gravity:** 5.55 **Colour:** Grey, greenish grey, yellowish; may turn to brownish-purple on exposure to light **Streak:** White **Lustre:** Resinous, adamantine; transparent to translucent

Chlorargyrite is a chloride of silver that forms in veins containing silver, where oxidation has occurred, especially in arid climates. It is often associated with native silver, galena, cerussite, pyromorphite, mimetite, wulfenite and malachite. It is not soluble in acids or water but can be dissolved in ammonia. Chlorargyrite melts in a flame.

▶ Chlorargyrite from Broken Hill, Yancowinna County, New South Wales, Australia.

Carnallite

Chemical composition: $KMgCl_3.6H_2O$ **Crystal system:** Orthorhombic **Mineral habit:** Rare pyramidal pseudo-hexagonal or tabular crystals; also massive and granular **Cleavage:** None **Fracture:** Conchoidal **Hardness:** 2½ **Specific gravity:** 1.60 **Colour:** Colourless, white, yellow, sometimes reddish due to iron inclusions **Streak:** White **Lustre:** Greasy; transparent to translucent

An evaporite mineral, carnallite forms with halite, gypsum, anhydrite, sylvite and polyhalite. Like many of the evaporite minerals, it is soluble in cold water and has a bitter, salty taste. When placed in a flame, carnallite melts with ease and colours the flame purple, due to the potassium in its chemical structure. As it is deliquescent, absorbing moisture from the air, it must be kept in an air-tight container.

▼ Carnallite from Beienrode, Göttingen, Lower Saxony, Germany.

Cryolite

Chemical composition: Na_3AlF_6 **Crystal system:** Monoclinic **Mineral habit:** Rare pseudocubic and prismatic crystals, commonly twinned and striated; usually granular and massive **Cleavage:** None **Fracture:** Uneven **Hardness:** 2½ **Specific gravity:** 2.97 **Colour:** Colourless, white, brownish, red, yellow **Streak:** White **Lustre:** Vitreous, greasy; transparent to translucent

Cryolite is an uncommon mineral formed in some igneous pegmatites, and can occur with fluorite, siderite, astrophyllite, riebeckite, zircon and topaz. It can be virtually invisible when placed in water. Cryolite melts with ease and colours a flame yellow, owing to sodium in its chemistry. It is soluble in sulphuric acid. Cryolite was once used as an ore of aluminium.

▶ Cryolite from Ivigtut, Arsuk Fjord, Greenland.

Boleite

Chemical composition: $Pb_{26}Ag_{10}Cu_{24}Cl_{62}(OH)_{48}\cdot 3H_2O$ **Crystal system:** Cubic **Mineral habit:** Pseudo-cubic crystals **Cleavage:** Perfect **Fracture:** Uneven **Hardness:** 3 to 3½ **Specific gravity:** 5.05 **Colour:** Deep blue **Streak:** Bluish **Lustre:** Vitreous, pearly; translucent

Boleite, a complex halide of lead, silver and copper, forms in lead and copper deposits when leaching by chloride-rich solutions has altered the original minerals. It can be found in clays with other halides, and is often associated with pseudoboleite, a very similar mineral with a slightly different chemical composition. Melting easily in a flame, boleite is soluble in nitric acid but not in water.

◀ Boleite from Boleo, Baja California, Mexico.

Diaboleite

Chemical composition: $Pb_2CuCl_2(OH)_4$ **Crystal system:** Tetragonal **Mineral habit:** Crystals tabular; also massive, as plates and grains. **Cleavage:** Perfect **Fracture:** Conchoidal **Hardness:** 2½ **Specific gravity:** 5.42 **Colour:** Deep blue **Streak:** Pale blue **Lustre:** Vitreous; transparent to translucent

Diaboleite forms in hydrothermal veins where primary minerals are altered by weathering or by fluids rising from depth. It is found with boleite, cerussite, linarite, hydrocerussite, caledonite and atacamite. Diaboleite can also occur with ores of manganese and in mining and smelting slags altered by contact with sea water.

▼ Diaboleite from Mammoth-Saint Anthony Mine, Tiger, Pinal Co., Arizona, USA.

Atacamite

Chemical composition: $Cu_2Cl(OH)_3$ **Crystal system:** Orthorhombic **Mineral habit:** Crystals prismatic and tabular, often striated and commonly twinned; also granular, massive, encrusting and fibrous. **Cleavage:** Perfect **Fracture:** Conchoidal **Hardness:** 3 to 3½ **Specific gravity:** 3.76 **Colour:** Bright to dark green **Streak:** Apple green **Lustre:** Adamantine to vitreous; transparent to translucent

▼ Atacamite from Caracoles, Sierra Gorda, Chile.

Atacamite commonly forms as a secondary mineral in the oxidized regions of copper deposits, especially in arid climates. It is usually associated with malachite, azurite, linarite, brochantite, cuprite and quartz. Atacamite can also occur around volcanic vents and fumaroles and in the crusts of material around deep seabed black smokers.

Fluorite

Chemical composition: CaF_2 **Crystal system:** Cubic **Mineral habit:** Crystals cubic, sometimes octahedral, commonly twinned; also massive, granular, botryoidal and banded **Cleavage:** Perfect **Fracture:** Subconchoidal **Hardness:** 4 **Specific gravity:** 3.18 **Colour:** Colourless, purple, green, blue, yellow, white, pink, brown and grey **Streak:** White **Lustre:** Vitreous; transparent to translucent

Fluorite is a common mineral in hydrothermal veins, where it can be found with calcite, quartz, barite, galena and sphalerite. It also forms around hot springs and as an accessory mineral in granite. The banded variety, Blue John, is used ornamentally, and is predominantly purple, white and colourless. The wealth of different colours in which fluorite can form makes it a popular material for mineral collectors. Fluorite is used as a flux in steel manufacture and refined for its fluorine content.

▶ Fluorite, St. Peter's Mine, Northumberland, England, UK.

Oxides and hydroxides

Oxide minerals are composed of various elements combined with oxygen. These minerals vary considerably in their appearance and properties. Some, such as the iron oxides, hematite and magnetite, are important metal ores. Corundum and its gemstone varieties, ruby and sapphire, are very hard and prized for their beauty. Hydroxides are formed by the combination of metallic elements with water and the hydroxyl ion (OH).

Oxides

Ice

Chemical composition: H_2O **Crystal system:** Hexagonal **Mineral habit:** Ice commonly forms as star-shaped crystals that are six-rayed, flattened and often twinned; also as hexagonal prisms and as aggregates of elongated lath-shaped crystals on the surface of water bodies; hailstones are small, globular masses of ice that can have a concentric internal structure **Cleavage:** None **Fracture:** Conchoidal **Hardness:** 1½ **Specific gravity:** 0.92 **Colour:** Colourless; also white when there are inclusions of gas bubbles; in thicker layers it may be pale blue to greenish blue **Streak:** Colourless **Lustre:** Vitreous; transparent

Ice is the solid phase of water that occurs below 0°C (32°F) at normal surface pressures. It forms on water surfaces, especially where water is not fast flowing, though in very cold conditions even waterfalls may freeze and become solid ice masses. Ice is often a meteorological phenomenon, occurring as snow, hail and hoar frost. If the temperature is below 0°C (32°F) for a prolonged period, large masses of ice can form and, in mountain and polar regions, glaciers may develop, existing for thousands of years.

▼ Ice crystals.

Spinel

Chemical composition: $MgAl_2O_4$ **Crystal system:** Cubic **Mineral habit:** Crystals octahedral, rarely cubic or dodecahedral, often twinned; also massive, compact and granular **Cleavage:** None **Fracture:** Conchoidal to uneven **Hardness:** 7½ to 8 **Specific gravity:** 3.60 to 4.10 **Colour:** Red, green, brown, black, blue **Streak:** White **Lustre:** Vitreous; transparent to opaque

Spinel is a widespread mineral, forming in many metamorphic rocks, including marble, gneiss and serpentinite. It is also found in some basic igneous rocks. In ultra-basic rocks such as peridotite, a chromium-bearing form of spinel occurs, which is thought to have formed deep in the Earth's mantle. Because of its hardness, spinel survives erosion and occurs in river gravels and sands as a placer deposit. It has an extremely high melting point and cannot be dissolved. Good specimens are prized as gemstones. Synthetic spinel has been created and has commercial applications.

▼ Spinel from Mogok, Myanmar.

Gahnite

Chemical composition: $ZnAl_2O_4$ **Crystal system:** Cubic **Mineral habit:** Crystals octahedral, rarely cubic or dodecahedral, often twinned; also massive, compact and granular **Cleavage:** Indistinct **Fracture:** Conchoidal to uneven **Hardness:** 7½ to 8 **Specific gravity:** 4.62 **Colour:** Bluish-black, greenish-black, yellowish, brown **Streak:** Grey **Lustre:** Vitreous to greasy; translucent to opaque

Closely related to spinel, gahnite contains zinc in place of spinel's magnesium. It forms in metamorphic rocks, especially schists and marbles, and in igneous pegmatites. Gahnite also occurs in replacement deposits and as an alteration product of sphalerite. Associated minerals include quartz, pyrite, chalcopyrite, corundum, staurolite and pyrrhotite. As with spinel, gahnite can be found in placer sands and gravels.

▲ Gahnite from Rhineland-Palatinate, Germany.

Zincite

Chemical composition: ZnO **Crystal system:** Hexagonal **Mineral habit:** Crystals pyramidal; also granular, massive and platy **Cleavage:** Perfect **Fracture:** Conchoidal **Hardness:** 4 **Specific gravity:** 5.64 to 5.68 **Colour:** Orange-yellow, deep yellow, dark red **Streak:** Orange-yellow **Lustre:** Subadamantine; transparent to translucent

Zincite forms in rocks that have been changed by contact metamorphism, and as a secondary mineral in altered zinc ore deposits. Associated minerals include franklinite, calcite, willemite, hemimorphite, sphalerite, smithsonite and tephroite. Zincite dissolves in hydrochloric acid but does not melt in a flame.

It is a rare mineral of interest to collectors. One of its best-known occurrences is at Franklin, New Jersey, USA. It has also been found in Spain, Poland, Italy, Namibia and Australia.

▼ Zincite from New Jersey, USA.

Franklinite

Chemical composition: $Zn^{2+}Fe^{3+}_2O_4$ **Crystal system:** Cubic **Mineral habit:** Crystals octahedral, often with rounded edges; also massive, granular and compact **Cleavage:** None **Fracture:** Uneven to subconchoidal **Hardness:** 5½ to 6 **Specific gravity:** 5.07 to 5.22 **Colour:** Black **Streak:** Black or reddish-brown **Lustre:** Metallic or dull; opaque

A member of the spinel group, franklinite forms in zinc deposits in metamorphosed limestones, with many other minerals including garnet, willemite, zincite, calcite and rhodonite. It is magnetic, a property that increases with heating. It does not melt in a flame, but dissolves in hydrochloric acid. Franklinite is a rare mineral. This mineral is mainly found in Franklin and Sterling, New Jersey, USA.

▼ Franklinite from Franklin Furnace, Sussex County, New Jersey, USA.

Cuprite

Chemical composition: Cu_2O **Crystal system:** Cubic **Mineral habit:** Crystals cubic, dodecahedral or octahedral; also massive, as grains, mats and hairlike structures **Cleavage:** Poor **Fracture:** Conchoidal to uneven **Hardness:** 3½ to 4 **Specific gravity:** 6.14 **Colour:** Dark red to brownish-red and black **Streak:** Brownish-red **Lustre:** Adamantine or submetallic to earthy; translucent to transparent or opaque

▶ Cuprite.

This widespread mineral forms in the parts of copper deposits that have been altered by oxidation. Here cuprite is associated with many minerals, including malachite, azurite, native copper, chalcocite and iron oxide. When exposed to the atmosphere, cuprite may become tarnished. It melts, colouring the flame green because of its copper content, and is soluble in concentrated acids. A hair-like variety called chalcotrichite is highly prized by mineral collectors.

Chromite

Chemical composition: $FeCr_2O_4$ **Crystal system:** Cubic **Mineral habit:** Crystals octahedral, rare; commonly massive and granular **Cleavage:** None **Fracture:** Uneven **Hardness:** 5½ **Specific gravity:** 4.50 to 4.80 **Colour:** Black **Streak:** Dark brown **Lustre:** Metallic; opaque

▼ Chromite from Verkhny Tagil, Sverdlovsk Oblast, Russia.

Chromite forms in ultrabasic and serpentinized igneous rocks, often occurring with magnetite, pyrrhotite, talc and pyroxenes. Certain placer sands and gravels contain chromite, and it occurs in meteorites and lunar basalts. It can be weakly magnetic and is insoluble in acids; when placed in a flame, it will not melt. This widespread mineral is the most important ore of the metal chromium, which is used as an alloy in steel and as a surface plating on metals and ceramics.

Magnetite

Chemical composition: Fe_3O_4 **Crystal system:** Cubic **Mineral habit:** Crystals octahedral or dodecahedral, striated, commonly twinned; also massive, granular and compact **Cleavage:** None **Fracture:** Uneven to subconchoidal **Hardness:** 5½ to 6½ **Specific gravity:** 5.17 **Colour:** Black **Streak:** Black **Lustre:** Metallic to dull; opaque

▼ Magnetite from Traversella, Piedmont, Italy.

Magnetite frequently occurs in magmatic rocks and in mineral veins rich in sulphides. It is also found as an important accessory mineral in some metamorphic rocks and as concentrations in placer sands. The name of this mineral reflects its magnetism; it attracts iron filings and deflects a compass needle. The latter property is so pronounced that a compass gives an incorrect reading in regions where magnetite is relatively abundant, as in basaltic and gabbroic rocks. Magnetite is a major ore of iron and was once called lodestone.

Ilmenite

Chemical composition: $FeTiO_3$ **Crystal system:** Trigonal **Mineral habit:** Crystals tabular, sometimes twinned; also granular, massive and lamellar **Cleavage:** None **Fracture:** Conchoidal to uneven **Hardness:** 5 to 6 **Specific gravity:** 4.68 to 4.76 **Colour:** Black **Streak:** Black **Lustre:** Metallic to dull; opaque

pyrrhotite. The concentration of many small water-worn grains of ilmenite and magnetite can colour placer sands black. Ilmenite has been detected in meteorites. When powdered, it dissolves in concentrated hydrochloric acid and exhibits weak magnetism if heated.

Ilmenite is found as an accessory mineral in a variety of igneous rocks, and in some metamorphic rocks. Associated minerals include hematite, magnetite, apatite, rutile and

▶ Ilmenite from Ilmen Mountains, Chelyabinsk Oblast, Russia.

Hematite

Chemical composition: Fe_2O_3 **Crystal system:** Trigonal **Mineral habit:** Crystals tabular, rhombohedral, pyramidal, rarely prismatic, often striated and twinned; usually massive, fibrous, botryoidal, reniform, compact, granular and sometimes stalactitic **Cleavage:** None **Fracture:** Uneven to subconchoidal **Hardness:** 5 to 6 **Specific gravity:** 5.26 **Colour:** Black, red **Streak:** Deep red to brownish red **Lustre:** Metallic to dull; opaque

A number of forms of hematite occur. When tabular crystals form in radiating groups, the name 'iron rose' is occasionally used; rounded masses with reniform habit are referred to as 'kidney ore', while 'specularite' is the form that occurs as aggregates of dark, metallic crystals. Hematite is a very important ore of iron; it can form in layered masses over 300 m (984 ft) thick. It is found in igneous rocks as an accessory mineral, in veins and especially as a replacement mineral. Hematite is a common alteration mineral in sedimentary rocks, colouring them red. When heated, hematite becomes magnetic, but does not melt. It dissolves in heated concentrated hydrochloric acid.

▲ Hematite from Cumbria, England, UK showing botryoidal form.

▲ Hematite from Cumbria, England, UK.

◄ Specularite from Cleator Moor, Cumbria, England, UK.

Chrysoberyl

Chemical composition: $BeAl_2O_4$ **Crystal system:** Orthorhombic **Mineral habit:** Crystals tabular, prismatic, striated, commonly twinned; also massive and granular. **Cleavage:** Distinct **Fracture:** Conchoidal to uneven **Hardness:** 8½ **Specific gravity:** 3.75 **Colour:** Various shades of green, yellow, brown, grey **Streak:** White **Lustre:** Vitreous, sometimes chatoyant; transparent to translucent

Chrysoberyl forms in a variety of metamorphic rocks including schist, gneiss and marble. It is also found in pegmatites and, being so hard, it resists erosion and occurs in river sands and placer deposits. Some green varieties show a colour change from green to red under fluorescent and incandescent light; these are known as alexandrite and have been used as gemstones. Chrysoberyl is insoluble and has an extremely high melting temperature.

▼ Chrysoberyl from Santa Teresa, Espírito Santo, Brazil showing twinning.

▼ Chrysoberyl variety alexandrite from Zimbabwe.

Cassiterite

Chemical composition: SnO_2 **Crystal system:** Tetragonal **Mineral habit:** Crystals prismatic or bipyramidal, commonly twinned; also granular and massive; rarely reniform, botryoidal or nodular **Cleavage:** Imperfect **Fracture:** Subconchoidal to uneven **Hardness:** 6 to 7 **Specific gravity:** 6.98 to 7.01 **Colour:** Brown to black, rarely colourless, red, greenish, yellowish or grey **Streak:** White, brownish, grey **Lustre:** Adamantine, vitreous; greasy when fractured; transparent to nearly opaque

Cassiterite forms in hydrothermal veins and rocks altered by metasomatism. It occurs in rocks affected by contact metamorphism and in granitic pegmatites, where it can be associated with tourmaline, quartz, wolframite, topaz, bismuth minerals, molybdenite and chalcopyrite. It is also found in placer deposits. Cassiterite has a very high melting temperature and cannot be easily dissolved in acid. This mineral is the most important ore of tin, which is used for tin plating and in the glass-making industry.

◀ Cassiterite pseudomorph from Wheal Coates, Cornwall, England, UK.

▶ Cassiterite from Panasqueira Mines, Covilhã, Castelo Branco, Portugal.

Pyrolusite

Chemical composition: MnO_2 **Crystal system:** Tetragonal **Mineral habit:** Crystals prismatic; usually massive, fibrous, columnar, compact, earthy, concretionary, stalactitic or dendritic **Cleavage:** Perfect **Fracture:** Uneven **Hardness:** 2 to 6½ **Specific gravity:** 5.04 to 5.08 **Colour:** Black to dark grey **Streak:** Black or bluish black **Lustre:** Metallic to dull; opaque

Pyrolusite forms from the alteration of manganese-rich minerals, often as a secondary mineral in veins. It also occurs in nodules on the deep seabed and is precipitated in bogs and lakes. The dendritic variety is found on bedding planes in sedimentary rocks and often resembles plant material. This mineral is soluble in hydrochloric acid, but does not melt if heated in a flame. When soft varieties are handled, dark, sooty marks are left on the hands. Pyrolusite is an important ore of manganese.

▲ Pyrolusite.

▼ Pyrolusite from Monte Albero, Elba, Italy.

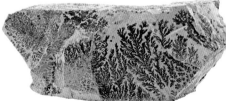

Corundum

Chemical composition: Al_2O_3 **Crystal system:** Trigonal **Mineral habit:** Crystals prismatic, tabular, rhombohedral, often striated, barrel-shaped and twinned; also granular, massive, bladed and compact **Cleavage:** None **Fracture:** Conchoidal to uneven **Hardness:** 9 **Specific gravity:** 3.98 to 4.10 **Colour:** Brownish, grey, blue, red, orange, green, yellow, purple, colourless **Streak:** White **Lustre:** Vitreous to adamantine; transparent to translucent

At point 9 on the Mohs scale of hardness, corundum is second only to diamond. It occurs in a variety of metamorphic rocks, including schist, gneiss and marble, and in granite and syenite. The red variety and the blue variety are highly prized as gemstones because of their colour, hardness, transparency and lustre. The colouring of the gemstone varieties of corundum is probably due to traces of other metals within the mineral. For example, ruby may be coloured by chromium and sapphire by iron and titanium. The granular, compact form of corundum is known as emery. Due to its hardness, corundum is used as an abrasive. Gem varieties are very often found in placer deposits, in alluvial sands and gravels; ruby is the red variety and the term sapphire is used for all other colours, although is best known as the blue variety.

▶ Corundum var. ruby from Mysore, India.

◀ Cut hexagonal corundum crystal from Zimbabwe.

◀ Massive sapphire from Mull, Scotland and small sapphire crystals from Australia.

▼ Rubies form Mysore, India.

▲ Hexagonal corundum crystal, from Zimbabwe.

Rutile

Chemical composition: TiO_2 **Crystal system:**
Tetragonal **Mineral habit:** Crystals prismatic,
often acicular, striated and twinned, rarely
pyramidal; also granular, massive and compact
Cleavage: Distinct **Fracture:** Conchoidal to
uneven **Hardness:** 6 to 6½ **Specific gravity:**
4.23 **Colour:** Reddish, brown, red, orange-
yellow, yellow, black **Streak:** Pale brown or
yellowish **Lustre:** Adamantine to submetallic;
transparent to translucent or opaque

▼ A prismatic rutile
from Lincoln Co.,
Georgia, USA.

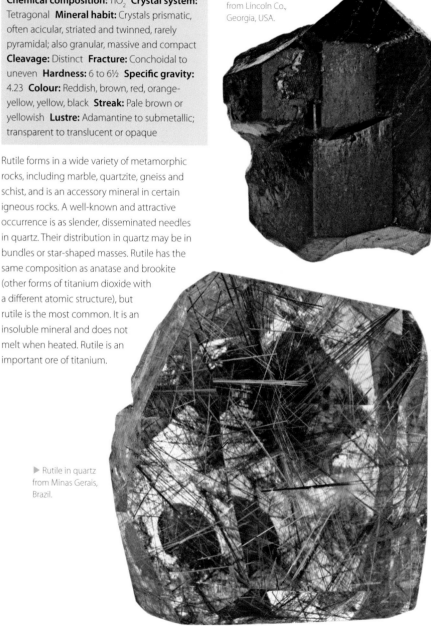

Rutile forms in a wide variety of metamorphic
rocks, including marble, quartzite, gneiss and
schist, and is an accessory mineral in certain
igneous rocks. A well-known and attractive
occurrence is as slender, disseminated needles
in quartz. Their distribution in quartz may be in
bundles or star-shaped masses. Rutile has the
same composition as anatase and brookite
(other forms of titanium dioxide with
a different atomic structure), but
rutile is the most common. It is an
insoluble mineral and does not
melt when heated. Rutile is an
important ore of titanium.

▶ Rutile in quartz
from Minas Gerais,
Brazil.

Uraninite

Chemical composition: UO_2 **Crystal system:** Cubic **Mineral habit:** Crystals cubic, octahedral, dodecahedral; also massive, granular, botryoidal and columnar; rarely fibrous **Cleavage:** Indistinct **Fracture:** Conchoidal to uneven **Hardness:** 5 to 6 **Specific gravity:** 10.63 to 10.95 **Colour:** Black, brownish, grey, greenish **Streak:** Black, brownish-black, grey, greenish **Lustre:** Submetallic or greasy; opaque

Uraninite forms in hydrothermal veins, granite and syenite pegmatites and sedimentary rocks. This mineral is highly radioactive and is the main source of uranium. The uranium in uraninite decays spontaneously and the ore often contains lead, which is a decay product. This process of decay has been used for the radiometric dating of rocks. Uraninite does not dissolve in hydrochloric acid, but is soluble in nitric, sulphuric and hydrofluoric acids. It will not melt when heated.

◀ Uraninite from Pribram, Czech Republic.

▶ Uraninite from Oberschlema, Ore Mountains, Germany.

Quartz

Chemical composition: SiO_2 **Crystal system:** Trigonal **Mineral habit:** Crystals prismatic, often pyramidal, striated, twinned; may be distorted; also massive, granular and concretionary **Cleavage:** None **Fracture:** Conchoidal to uneven **Hardness:** 7 **Specific gravity:** 2.65 **Colour:** Colourless, white, grey, yellow, green, purple, pink, brown, black **Streak:** White **Lustre:** Vitreous; transparent to translucent

Quartz is a very common mineral that forms as an essential component of many igneous rocks, especially those of acidic composition, such as granite. It also occurs in a variety of metamorphic rocks, such as schist and gneiss, and as detrital grains in sandstones and a wide range of sedimentary rocks. Quartz is also common in mineral veins, with metallic ore minerals. Many references list quartz as a silicate, but correctly it is an oxide of silicon. Because of its hardness and colour varieties, quartz is widely used as a gemstone. The colourless, transparent form, rock crystal, often contains small needles of rutile. When white, the term milky quartz is used. Amethyst is purple-coloured quartz; this transparent to translucent variety is an attractive gemstone. Smoky quartz can be pale brown to almost black in colour. The very dark form is called morion. Citrine is a yellowish to red-brown colour. An attractive pale pink to deep rose pink colour form is called rose quartz; this very rarely occurs as crystals, usually exhibiting a massive habit. There are a number of cryptocrystalline varieties of quartz in which the crystals are minute and poorly formed. Chalcedony is a type of cryptocrystalline quartz with a micro-fibrous crystal structure; it often occurs as botryoidal masses and concretions, exhibiting a variety of colours ranging from white to grey and blue. Banded forms are known as agate; this frequently occurs as concentric, banded masses in vugs and amygdales in basaltic lava. These tend to weather out of lava, as they are harder than the rock in which they are enclosed. Agate is much used in jewellery. Onyx is a variety of agate with straight, parallel bands. Chrysoprase is greenish coloured chalcedony rich in nickel, while carnelian is a translucent, reddish-orange form of chalcedony. A red coloured, opaque, cryptocrystalline type of quartz is called jasper. Chert and flint are terms applied to discrete, nodular masses of cryptocrystalline quartz, usually found in limestones. All these colour variations are probably the result of minute amounts of different elements such as manganese and iron in the crystal structure and tiny inclusions of other minerals.

◀ Milky quartz from India.

▼ Amethyst from Idar-Oberstein, Rhineland-Palatinate, Germany.

▶ Quartz var. citrine from Brazil.

▶ Rock crystal from Maderanertal, Uri, Switzerland.

▲ Smoky quartz from Switzerland.

▶ Quartz var. agate from Laguna District, Mexico.

◀ Rose quartz from Minas Gerais, Brazil.

▶ Chrysoprase from Marlborough, Queensland, Australia.

▼ Carnelian from India.

◀ Quartz var. Jasper.

Opal

Chemical composition: $SiO_2.nH_2O$ **Crystal system:** None **Mineral habit:** Massive, globular, compact, earthy, reniform, botryoidal, stalactitic and nodular **Cleavage:** None **Fracture:** Conchoidal **Hardness:** 5½ to 6½ **Specific gravity:** 1.90 to 2.30 **Colour:** Colourless, white, bluish; yellowish, reddish-brown, brown, greenish, grey, black, occasionally with very colourful internal features **Streak:** White **Lustre:** Vitreous to resinous, pearly and waxy; transparent to nearly opaque

Opal can form in a great variety of geological situations, at or near the surface, especially around hot springs, where it is precipitated from silica-rich water. A number of gemstone varieties of opal are recognized. Precious opal is widely used in jewellery, because of the play of bright colours that can be seen within, thought to be the result of its internal structure of minute silica spheres, which refract light. When heated, and the water in its chemical structure is removed, the brilliant colours can disappear. Fire opal owes its brownish red colouring to the presence of iron in its structure. It has similar properties to precious opal. Wood opal is a form of fossilized wood, where the plant material has been replaced by opal. It commonly retains the concentric growth bands of the wood, with a brownish to black colouring.

▲ Fire opal from Australia.

▲ Precious opal of the variety boulder opal, from Australia.

▶ Wood opal from Hobart, Tasmania, Australia.

Columbite series

Chemical composition: $(Fe,Mn)(Nb,Ta)_2O_6$
Crystal system: Orthorhombic **Mineral habit:**
Crystals tabular, prismatic, pyramidal, often
twinned; also massive and compact **Cleavage:**
Distinct **Fracture:** Subconchoidal to uneven
Hardness: 6 **Specific gravity:** 5.20 to 6.65
Colour: Black or pale green **Streak:** Black to
brownish black **Lustre:** Submetallic; opaque

▲ Columbite from
Portland County, USA.

Strictly, this is two minerals, ferrocolumbite and
manganocolumbite, depending on whether the
material is richer in iron or manganese. These
minerals form in pegmatites and in placer deposits
derived from these by weathering and erosion.
They are associated with a variety of minerals,
including quartz, tourmaline, beryl, spodumene,
micas, cassiterite and feldspars. The columbites
have a very similar chemical composition to
ferrotantalite and manganotantalite, all four

minerals forming a series. The tantalites have a
much higher specific gravity, around 8.00, and are
the main source of tantalum, a metal valued for
its resistance to corrosion. Together, these similar
minerals from the same geological environments
are often termed 'coltan'.

Pyrochlore group

Chemical composition: $(Na,Ca,U)_2(Nb,Ta,Ti)_2$
$O_6(OH,F)$ **Crystal system:** Cubic **Mineral
habit:** Crystals octahedral; also granular
and earthy **Cleavage:** Distinct **Fracture:**
Subconchoidal to uneven **Hardness:** 5 to 5½
Specific gravity: 4.45 to 4.90 **Colour:**
Yellowish-brown, reddish-brown, brown,
black **Streak:** Brownish to yellowish **Lustre:**
Vitreous or resinous; translucent to opaque

◀ Pyrochlore
from Vishnevye
Mountains,
Chelyabinsk Oblast,
Russia.

Not strictly a mineral name, the term pyrochlore
refers to an entire group of very similar minerals
with the same structure but different chemical
compositions. All the elements in the formula
listed above can substitute for each other and
different combinations lead to different scientific
mineral names such as fluornatromicrolite and
oxycalciopyrochlore, the names reflecting the
specific chemical composition. The pyrochlores

form in many geological environments, but large
crystals are best known in igneous rocks, including
pegmatites, syenites and carbonatites, in association
with fluorite, feldspar, tantalite, beryl, astrophyllite,
tourmaline, topaz, spessartine and zircon. Pyrochlore
minerals are important ores of niobium and are
mined in Brazil, Canada, Russia, Zaire and Nigeria.

Perovskite

Chemical composition: $CaTiO_3$ **Crystal system:** Orthorhombic **Mineral habit:** Crystals pseudocubic and pseudo-octahedral, striated, sometimes twinned; also massive, granular, reniform and lamellar **Cleavage:** Imperfect **Fracture:** Subconchoidal to uneven **Hardness:** 5½ **Specific gravity:** 3.98 to 4.26 **Colour:** Dark brown, black, amber, yellow **Streak:** Pale grey **Lustre:** Adamantine, metallic, dull; transparent to opaque

▼ Perovskite from Achmatovsk Mine, Nazyamskie Mountains, Zlatoust, Chelyabinsk Oblast, Russia.

Perovskite forms in basic and ultrabasic igneous rocks; and is an important component of deep crustal rocks. It is found in kimberlite as an accessory mineral. Perovskite also occurs in certain metamorphic rocks, especially chlorite- and talc-rich schists and limestone altered by contact metamorphism. This mineral will not melt in a flame, but can be dissolved in heated sulphuric acid.

Hubnerite

Chemical composition: $MnWO_4$ **Crystal system:** Monoclinic **Mineral habit:** Prismatic or tabular, striated and often twinned; frequently in parallel or radiating groups **Cleavage:** Perfect **Fracture:** Uneven **Hardness:** 4 to 4½ **Specific gravity:** 7.12 to 7.18 **Colour:** Reddish-brown, yellowish-brown, sometimes with an iridescent tarnish **Streak:** Reddish brown, yellowish, greenish grey **Lustre:** Submetallic or resinous; transparent to translucent

◄ Hubnerite from Pasto Bueno, Peru.

Hubnerite is an end-member of the hubnerite-ferberite series, in which hubnerite contains manganese, while ferberite contains iron. Ferberite has similar features to hubnerite, but its density is slightly higher at 7.51 and it is black or dark brown in colour, with a brownish black to black streak. Ferberite tends to form as wedge-shaped crystals and can have a massive habit. It is an opaque mineral with a submetallic lustre. Both minerals occur in hydrothermal veins associated with granite. Undifferentiated members of the series are collectively called wolframite. Members of the hubnerite-ferberite series are ores of tungsten and are insoluble in acids.

Hydroxides

Gibbsite

Chemical composition: $Al(OH)_3$ **Crystal system:** Monoclinic **Mineral habit:** Crystals tabular; also massive, concretionary, stalactitic and as coatings and crusts **Cleavage:** Perfect **Fracture:** Uneven **Hardness:** 2½ to 3 **Specific gravity:** 2.38 to 2.42 **Colour:** White, greyish, greenish, reddish **Streak:** White **Lustre:** Vitreous to pearly; translucent to transparent

▼ Gibbsite from Wismar, Upper Demerara-Berbice Region, Guyana.

Gibbsite is a widespread mineral that occurs in the alteration zones of aluminium-rich mineral deposits in bauxites and laterites. It also forms in hydrothermal veins, associated with igneous rocks. In bauxites it is an important ore of aluminium and makes up over a third of the rock. Associated minerals can include goethite, kaolinite, diaspore and corundum.

Brucite

Chemical composition: $Mg(OH)_2$ **Crystal system:** Trigonal **Mineral habit:** Crystals tabular; usually massive, foliated, fibrous, granular and scaly **Cleavage:** Perfect **Fracture:** Uneven **Hardness:** 2½ to 3 **Specific gravity:** 2.39 **Colour:** White, blue, greenish, grey; manganese-rich forms can be yellow or brown **Streak:** White **Lustre:** Pearly, waxy, vitreous; transparent

▲ Brucite from Asbest, Sverdlovsk Oblast, Russia.

Brucite forms in marbles and serpentinites and also in schists, producing an attractive green or bluish veining in marble. It occurs with a variety of minerals, including dolomite, calcite, aragonite, talc, chrysotile, magnesite and artinite. Brucite dissolves in acids, but will not melt when heated in a flame.

Goethite

Chemical composition: FeO(OH) **Crystal system:** Orthorhombic **Mineral habit:** Crystals prismatic, striated, acicular, thin tabular; usually massive, botryoidal, concretionary, oolitic, stalactitic and earthy **Cleavage:** Perfect **Fracture:** Uneven **Hardness:** 5 to 5½ **Specific gravity:** 4.27 to 4.29 **Colour:** Brownish-black, reddish-brown, yellowish-brown; when earthy, yellowish; rarely with an iridescent tarnish **Streak:** Orange to brownish-yellow **Lustre:** Metallic to silky or dull; opaque

▲ Goethite from Botallack Mine, St. Just, Cornwall, England, UK.

Goethite occurs in any geological environment where iron-bearing minerals have been altered by oxidation. These minerals include magnetite, pyrite, siderite, hematite, and chalcopyrite. It can be found in clay deposits, bogs and laterites and occasionally in hydrothermal veins. Goethite is soluble in hydrochloric acid and may become magnetic when heated. The variety name limonite is used for yellow earthy goethite and for mixtures of iron oxides and hydroxides that have not been analyzed as specific minerals.

Manganite

Chemical composition: MnO(OH) **Crystal system:** Monoclinic **Mineral habit:** Crystals prismatic, striated, often twinned, frequently in bundles; also massive, fibrous, columnar, granular, stalactitic and concretionary **Cleavage:** Perfect **Fracture:** Uneven **Hardness:** 4 **Specific gravity:** 4.29 to 4.34 **Colour:** Dark grey to black **Streak:** Red-brown to black **Lustre:** Submetallic to dull; opaque

Manganite forms in hydrothermal veins, bogs, lake sediments and in shallow marine environments. It occurs where weathering of clay and laterite takes place and where underground waters circulate. In certain situations manganite changes to pyrolusite without its form being altered. It is found with a number of other minerals, including pyrolusite, siderite, barite, calcite and goethite.

▶ Manganite from Ilfeld, Thuringia, Germany.

Carbonates, nitrates and borates

Minerals in the carbonate group are chemical compounds in which metals or semi-metals are combined with the carbonate ion (CO_3). Many members of this group display bright colours. Carbonate minerals often occur as well-formed crystals with a rhombohedral habit. When dilute hydrochloric acid is applied to them, a strong effervescent reaction is produced. Nitrates are compounds where the nitrate ion (NO_2) has combined with metallic elements. Borates are formed by metallic elements combining with the borate ion (BO_3).

Carbonates

Aragonite

Chemical composition: $CaCO_3$ **Crystal system:** Orthorhombic **Mineral habit:** Crystals acicular, often twinned and striated; also radiating, columnar, fibrous, coralline, stalactitic **Cleavage:** Distinct **Fracture:** Subconchoidal **Hardness:** 3½ to 4 **Specific gravity:** 2.95 **Colour:** Colourless, white, grey; pale red, brown, blue, green or yellowish, due to impurities **Streak:** White **Lustre:** Vitreous to resinous; transparent to translucent

Aragonite is a widespread mineral that usually occurs in limestone areas, and also near hot springs. It is found in sedimentary and metamorphic rocks and in the altered parts of ore deposits. This mineral can form biologically, especially in the shells of some molluscs. Chemically, it is the same as the most common carbonate mineral, calcite, and under some unusual conditions it can change into this mineral. The coralline form is called 'flos ferri'. Aragonite dissolves in weak hydrochloric acid with strong effervescence.

◀ Aragonite from Eisenerz, Styria, Austria.

▶ Aragonite from Eskett Mine, Copeland, Cumbria, England, UK.

Calcite

Chemical composition: $CaCO_3$ **Crystal system:** Trigonal **Mineral habit:** Crystals rhombohedral or scalenohedal, often twinned; also massive, fibrous, granular, stalactitic **Cleavage:** Perfect **Fracture:** Conchoidal **Hardness:** 3 **Specific gravity:** 2.71 **Colour:** Colourless, white; also greenish, grey, yellowish, brown, blue or almost black when coloured by impurities **Streak:** White or greyish **Lustre:** Vitreous to pearly; transparent to translucent

▶ Calcite from Millclose Mine, Derbyshire, England, UK.

This very common and widespread mineral occurs in many geological situations. It is the main mineral in limestone and marble and is found in hydrothermal veins together with metallic ores. Calcite forms stalactites and stalagmites in caves and is often present in concretions and geodes. It also occurs in some igneous rocks, notably carbonatites, which are made up mainly of calcite and dolomite. A number of the crystalline shapes are informally named. Nail-head calcite has flattened crystal shapes on smaller columnar structures, while the dog's tooth variety consists of groups of pyramidal crystals. The term Iceland spar is used for very pure, transparent crystals that show double refraction, where an object, such as a thin wire or some writing, placed below the calcite crystal will appear doubled when viewed through it. Calcite effervesces strongly with weak hydrochloric acid.

◀ Calcite from Gillfoot Park Mine, Egremont, Cumbria, England, UK.

▼ Nail-head calcite from Alston Moor, Cumbria, England, UK.

◀ Iceland spar, showing double refraction.

Dolomite

Chemical composition: $CaMg(CO_3)_2$ **Crystal system:** Trigonal **Mineral habit:** Crystals rhombohedral, often with curved faces, which can be 'saddle-shaped', prismatic, sometimes octahedral or tabular; also massive, granular **Cleavage:** Perfect **Fracture:** Subconchoidal **Hardness:** 3½ to 4 **Specific gravity:** 2.84 to 2.86 **Colour:** Colourless, white, greyish, brownish, pinkish, greenish **Streak:** White **Lustre:** Vitreous to pearly; transparent to translucent

Dolomite is a common mineral in many sedimentary rocks, especially dolostones, which are limestones composed mainly of this mineral. These are often associated with shallow marine deposition and can be found in some evaporite sequences. Dolomite also occurs in hydrothermal veins with ore minerals such as galena and sphalerite. Altered igneous rocks and serpentinites may also contain dolomite. This widespread mineral dissolves quickly in heated hydrochloric acid.

▼ Dolomite from Eugui, Navarra, Spain.

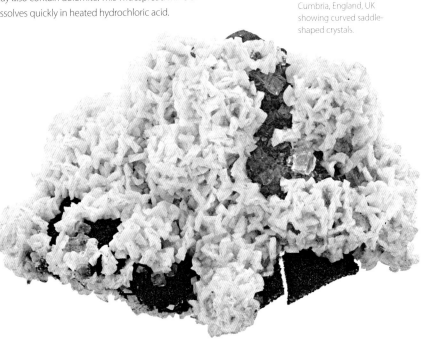

▼ Dolomite from Florence Mine, Egremont, Cumbria, England, UK showing curved saddle-shaped crystals.

Rhodochrosite

Chemical composition: $MnCO_3$ **Crystal system:** Trigonal **Mineral habit:** Crystals rhombohedral, sometimes tabular, prismatic; usually massive, granular, botryoidal or stalactitic **Cleavage:** Perfect **Fracture:** Conchoidal to uneven **Hardness:** 3½ to 4 **Specific gravity:** 3.70 **Colour:** Pale pink, deep rose-red, brown, yellowish, grey **Streak:** White **Lustre:** Vitreous to pearly; transparent to translucent

▼ Rhodochrosite from N'Chwaning Mines, Kuruman, Northern Cape, South Africa.

This beautifully coloured mineral occurs in hydrothermal mineral veins, manganese-rich deposits and where rocks have been altered by metasomatism. Rhodochrosite is associated with dolomite, calcite, siderite, fluorite, barite, sphalerite, quartz and other hydrothermal vein minerals, and in metamorphic rocks with garnet, rhodonite and alabandite. Even though it is a very soft mineral, easily scratched, banded forms are used ornamentally. Rhodochrosite dissolves in heated hydrochloric acid.

▼ Rhodochrosite from Capillitas Mining District, Catamarca Province, Argentina.

Ankerite

Chemical composition: $Ca(Fe^{2+},Mg)(CO_3)_2$
Crystal system: Trigonal **Mineral habit:**
Crystals rhombohedral, often twinned; also
massive, granular **Cleavage:** Perfect **Fracture:**
Subconchoidal **Hardness:** 3½ to 4 **Specific
gravity:** 2.93 to 3.10 **Colour:** White, grey,
brown, yellowish-brown **Streak:** White **Lustre:**
Vitreous to pearly; translucent

▼ Ankerite from
Boltsburn Mine,
Weardale, County
Durham, England, UK.

Ankerite forms in hydrothermal mineral veins with
a variety of sulphide ore minerals, and with siderite,
dolomite and gold, especially where these vein
deposits have been hydrothermally altered. It also
occurs where iron-rich rocks have been changed by
metamorphism. Ankerite is soluble in hydrochloric
acid. It is a fairly widespread mineral.

Smithsonite

Chemical composition: $ZnCO_3$ **Crystal
system:** Trigonal **Mineral habit:** Crystals
rhombohedral, sometimes scalenohedral,
often with curved faces; usually botryoidal,
reniform, stalactitic; also massive, granular
and as crusts and earthy masses. **Cleavage:**
Perfect **Fracture:** Subconchoidal to
uneven **Hardness:** 4 to 4½ **Specific
gravity:** 4.42 to 4.44 **Colour:** White, grey;
yellowish, brown, greenish, blue, pink **Streak:**
White **Lustre:** Vitreous to pearly or dull;
transparent to translucent

Smithsonite forms as a widespread secondary
mineral in copper and zinc ore deposits, when
alteration and oxidation have taken place. It can
also occur where limestones have been replaced
by zinc ores. It is found with a variety of other
minerals, including cerussite, azurite, malachite,
aurichalcite, willemite, hemimorphite and
anglesite. There has been confusion between the
blue varieties of smithsonite and blue-coloured
hemimorphite, but smithsonite is denser and
slightly less hard than hemimorphite. Smithsonite
dissolves in concentrated hydrochloric acid.

◀ Smithsonite from
Farnberry Mine,
Alston Moor, Cumbria,
England, UK.

Siderite

Chemical composition: $FeCO_3$ **Crystal system:** Trigonal **Mineral habit:** Crystals rhombohedral, tabular, prismatic, scalenohedral, often with curved faces, twinned; also botryoidal, oolitic, massive, granular **Cleavage:** Perfect **Fracture:** Uneven to conchoidal **Hardness:** 3½ to 4½ **Specific gravity:** 3.96 **Colour:** Yellowish, grey, greenish, brown, brownish-black **Streak:** White **Lustre:** Vitreous, pearly to silky or translucent

Siderite forms in a variety of rocks, especially those of sedimentary origin, and in mineral ore veins, associated with fluorite, quartz, barite, pyrite, galena and calcite. It occurs in sandstone and shale, and in rounded concretions, and has been

▼ Siderite from Drakewalls Mine, Calstock, Cornwall, England, UK.

found in pegmatites and syenite. When heated, siderite becomes magnetic. It dissolves in heated hydrochloric acid, with effervescence.

Strontianite

Chemical composition: $SrCO_3$ **Crystal system:** Orthorhombic **Mineral habit:** Crystals prismatic, acicular; usually massive, fibrous, concretionary or granular **Cleavage:** Perfect **Fracture:** Subconchoidal to uneven **Hardness:** 3½ **Specific gravity:** 3.74 to 3.78 **Colour:** Colourless, white, grey, yellowish, brownish, reddish, greenish **Streak:** White **Lustre:** Vitreous to resinous; transparent to translucent

Strontianite is found in hydrothermal veins and in geodes and veins in sedimentary limestones, chalk and marl, with harmotome, calcite, celestine and barite. It also occurs in carbonatites. Strontianite is a widespread ore of the metal strontium, the main uses of which are in specialized glass and providing the brilliant red colour in flares, fireworks and paint. This red colour is produced when the mineral is heated in a flame. Strontianite dissolves in hydrochloric acid.

▲ Strontianite from Whitesmith Mine, Strontian, Scotland, UK.

Witherite

Chemical composition: $BaCO_3$ **Crystal system:** Orthorhombic **Mineral habit:** Crystals tabular, prismatic, twinned, forming pseudohexagonal dipyramids, often striated; also massive, granular, fibrous, columnar **Cleavage:** Distinct **Fracture:** Uneven **Hardness:** 3 to 3½ **Specific gravity:** 4.29 **Colour:** Colourless, white; also grey, brownish, yellowish, greenish **Streak:** White **Lustre:** Vitreous to resinous; transparent to translucent

Witherite occurs in hydrothermal ore veins with many other minerals, including calcite, fluorite, galena and barite. It can be formed by the alteration of barite and its barium content gives the mineral a relatively high specific gravity. Witherite dissolves in hydrochloric acid, with effervescence, and when placed under ultraviolet light, it exhibits a bluish fluorescence. It melts when heated, turning the flame greenish-yellow. Locations in which witherite occurs include Northumberland, Durham and Cumbria (England), Illinois and California (USA), and Russia.

▶ Witherite from Nentsberry Haggs Mine, Alston Moor, Cumbria, England, UK.

Magnesite

Chemical composition: $MgCO_3$ **Crystal
system:** Trigonal **Mineral habit:** Rare crystals
rhombohedral, sometimes prismatic or
tabular; usually massive, compact, granular,
fibrous, chalky **Cleavage:** Perfect **Fracture:**
Conchoidal **Hardness:** 3½ to 4½ **Specific
gravity:** 2.98 to 3.02 **Colour:** White, grey,
brownish, yellowish **Streak:** White **Lustre:**
Vitreous to dull; transparent to translucent

▲ Magnesite from
Snarum, Buskerud,
Norway.

Magnesite is formed by the alteration of rocks rich
in magnesium, such as peridotite, in metamorphic
rocks like serpentinite and in some sedimentary
rocks. It also occurs in hydrothermal mineral veins.
Magnesite is often associated with dolomite,
calcite and talc. This mineral is an important and
widespread ore of magnesium, which is used
in the manufacture of heat-resistant bricks and
fertilizer and is alloyed with aluminium and zinc
for making car and aircraft bodies. Magnesite
effervesces in heated hydrochloric acid.

Artinite

Chemical composition: $Mg_2(CO_3)(OH)_2.3H_2O$
Crystal system: Monoclinic **Mineral habit:**
Acicular crystals in crusts and sprays; also as
fibrous and spherical masses **Cleavage:** Perfect
Fracture: Uneven **Hardness:** 2½ **Specific
gravity:** 2.01 to 2.03 **Colour:** White **Streak:**
White **Lustre:** Vitreous or silky; transparent

▲ Artinite from
San Benito County,
California, USA.

Artinite is a very brittle mineral that occurs as
coatings and crusts in serpentinized ultrabasic
rocks. It is found with a variety of minerals,
including chrysotile, hydromagnesite, brucite,
aragonite, magnesite, calcite and dolomite.
Artinite does not melt in a flame, but gives off
carbon dioxide and water. It dissolves easily in
dilute acids.

Cerussite

Chemical composition: $PbCO_3$ **Crystal system:** Orthorhombic **Mineral habit:** Crystals tabular, rarely acicular, often striated and twinned, frequently as straw-like masses; also massive, granular, sometimes fibrous **Cleavage:** Distinct **Fracture:** Conchoidal **Hardness:** 3 to 3½ **Specific gravity:** 6.53 to 6.57 **Colour:** Colourless, white, grey; bluish, greenish **Streak:** White **Lustre:** Adamantine, vitreous, pearly, resinous; transparent to translucent

Cerussite forms in mineral veins where metallic ores have been altered, often by oxidation. It occurs widely in lead deposits with galena, pyromorphite, smithsonite, malachite, azurite and anglesite. Because of its lead content, cerussite has a very high specific gravity for a carbonate mineral. It will not dissolve in hydrochloric acid, but is soluble in nitric acid, with effervescence. Cerussite may show yellow fluorescence under ultraviolet light.

▲ Cerussite from Broken Hill, New South Wales, Australia.

▶ Straw-like cerussite from Cumbria, England, UK, nick-named 'Jackstraw cerussite'.

Malachite

Chemical composition: $Cu_2CO_3(OH)_2$ **Crystal system:** Monoclinic **Mineral habit:** Rare crystals prismatic, acicular, small, commonly twinned; usually massive, botryoidal, stalactitic, banded or fibrous **Cleavage:** Perfect **Fracture:** Subconchoidal to uneven **Hardness:** 3½ to 4 **Specific gravity:** 3.60 to 4.05 **Colour:** Bright green to dark green **Streak:** Pale green **Lustre:** Vitreous to adamantine, silky or dull; transparent to opaque

▼ Botryoidal malachite from Zimbabwe.

Malachite forms in the altered parts of copper deposits, where minerals such as chalcopyrite and chalcocite have been oxidized. It is found with a number of other secondary minerals, including azurite, calcite and goethite. Malachite is a soft mineral, easily marked with a knife blade, but is nevertheless much used for decorative purposes, mainly because of its fine colour and banded structure. It is easily worked into different shapes and takes a good polish. Malachite dissolves in concentrated acids. When held in a flame, it melts and colours the flame green because of its copper content.

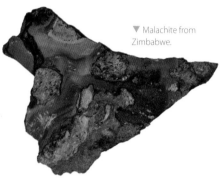

▼ Malachite from Zimbabwe.

▲ Banded malachite from Bwana M'Kubwa Mine, Ndola, Copperbelt Province, Zambia.

Azurite

Chemical composition: $Cu_3(CO_3)_2(OH)_2$
Crystal system: Monoclinic   **Mineral habit:**
Crystals tabular, prismatic, rhombohedral: also
massive, concretionary, stalactitic, earthy
Cleavage: Perfect   **Fracture:** Conchoidal
Hardness: 3½ to 4   **Specific gravity:** 3.77
Colour: Light to dark blue   **Streak:** Pale blue
Lustre: Vitreous or dull; transparent to opaque

▼ Azurite from
Boomerang Mine,
Mount Isa, Queensland,
Australia.

Azurite occurs in the parts of copper veins
that have been altered by oxidation and
weathering. It is associated with malachite,
chalcocite, chrysocolla and other secondary
copper minerals. Azurite is less stable than
malachite and can be replaced by that mineral.
In the past it has been used extensively as
a blue pigment, but its instability and pale
colour when powdered make it unsuitable for
this use. Because of its strong blue colour, it is sometimes used decoratively, even though it is easily scratched. Azurite is soluble in hydrochloric acid and is easily melted in a flame, turning black.

Aurichalcite

Chemical composition: $(Zn,Cu)_5(CO_3)_2(OH)_6$
Crystal system: Monoclinic   **Mineral habit:**
Crystals acicular; also as tufted aggregates, and
encrustations   **Cleavage:** Perfect   **Fracture:**
Uneven   **Hardness:** 1 to 2   **Specific gravity:**
3.96   **Colour:** Pale green, greenish-blue, sky-
blue   **Streak:** Pale blue-green   **Lustre:** Silky to
pearly; transparent

Aurichalcite forms as a secondary mineral in
veins containing copper and zinc ore minerals.
It is found worldwide, with azurite, malachite,
hydrozincite, smithsonite, cerussite, cassiterite
and hemimorphite. Aurichalcite will not melt, but
colours a flame green due to its copper content.
It dissolves in dilute hydrochloric acid, liberating
carbon dioxide.

▲ Aurichalcite
from 79 Mine, Gila
Co., Arizona, USA.

Nitrates

Nitratine

Chemical composition: $NaNO_3$ **Crystal system:** Trigonal **Mineral habit:** Crystals rhombohedral, often twinned; usually massive, granular, encrusting **Cleavage:** Perfect **Fracture:** Conchoidal **Hardness:** 1½ to 2 **Specific gravity:** 2.26 **Colour:** Colourless, white; pinkish, grey, yellowish, brown **Streak:** White **Lustre:** Vitreous; transparent

▼ Nitratine from Tarapacá, Chile.

In arid regions, nitratine occurs as an efflorescence on the surface and can be found with gypsum, often occurring over wide areas. It is readily soluble in water (hence its formation in arid areas) and melts easily in a flame, which it colours yellow. This mineral has also been called nitratite.

Borates

Boracite

Chemical composition: $Mg_3(B_7O_{13})Cl$ **Crystal system:** Orthorhombic **Mineral habit:** Crystals cubic, tetrahedral, dodecahedral, pseudo-octahedral; also fibrous, granular **Cleavage:** None **Fracture:** Conchoidal to uneven **Hardness:** 7 to 7½ **Specific gravity:** 2.91 to 3.10 **Colour:** Colourless, white, grey, green, yellow **Streak:** White **Lustre:** Vitreous; transparent to translucent

Boracite is a rare mineral that forms in bedded evaporite sequences with halite, sylvite, gypsum, anhydrite and carnallite. Unusually for an evaporite mineral, it has a high hardness and cannot be scratched with a steel blade. Boracite is soluble in hot hydrochloric acid, and produces the characteristic green colour associated with boron when placed in a flame. Occurrences include North Yorkshire (England), Louisiana and California (USA), France, Germany and Poland.

▶ Boracite from Boulby Mine, Loftus, North Yorkshire, England, UK.

Ulexite

Chemical composition: $NaCaB_5O_6(OH)_6.5H_2O$
Crystal system: Triclinic **Mineral habit:**
Crystals acicular; usually as nodules, crusts or
fibrous masses **Cleavage:** Perfect **Fracture:**
Uneven **Hardness:** 2½ **Specific gravity:** 1.95
Colour: Colourless, white **Streak:** White
Lustre: Vitreous, silky; transparent to translucent

▼ Ulexite from Bigadiç
Mine, Bigadiç, Balikesir
Province, Turkey.

Ulexite forms in evaporite deposits in playa
regions, associated with other evaporite minerals
such as glauberite, anhydrite, colemanite, trona,
calcite, gypsum, borax and halite. The structure
of parallel crystal fibres within ulexite produces a
fibre-optic effect, allowing light to run down the
mineral's long axis by internal reflection. This has
led to its being referred to as 'television stone'.
It is soluble in hot water and melts in a flame,
colouring it yellow. Ulexite is found in California
and Nevada (USA), Canada, Argentina, Peru, Chile,
Kazakhstan and Turkey.

Colemanite

Chemical composition: $Ca_2B_3O_4(OH)_3.H_2O$
Crystal system: Monoclinic **Mineral habit:**
Crystals prismatic, pseudorhombohedral; also
massive, granular and in geodes **Cleavage:**
Perfect **Fracture:** Subconchoidal to uneven
Hardness: 4½ **Specific gravity:** 2.42
Colour: Colourless, white, yellowish, grey,
brown **Streak:** White **Lustre:** Vitreous;
transparent to translucent

Colemanite occurs in arid regions in playa
lake deposits and other evaporite formations,
probably from the secondary alteration of borax
and ulexite, together with kernite, gypsum,
calcite, celestine and calcite. In some regions it is
mined as an ore of boron. This is added to metals
to improve their conductivity and is also used to
toughen glass, and in the manufacture of glazes.
Colemanite dissolves in heated hydrochloric acid
and melts in a flame, colouring it green because
of its boron content.

▶ Colemanite from
Boron, Kern Co.,
California, USA.

Sulphates, chromates, molybdates and tungstates

Minerals belonging to the sulphate group are formed when the sulphate anion (SO_4) combines with a metallic element, or elements, to form a compound. Sulphate minerals are quite common and many form as evaporites. Chromates are compounds of metallic elements with the chromate anion (CrO_4). These are quite rare minerals. Tungstates and molybdates form when metals combine with the tungstate anion (WO_4) or the molybdate anion (MoO_4) respectively.

Sulphates

Gypsum

Chemical composition: $CaSO_4.2H_2O$
Crystal system: Monoclinic **Mineral habit:** Crystals tabular, often diamond-shaped; sometimes prismatic, acicular, lenticular, commonly twinned, striated; also massive, fibrous, granular, concretionary **Cleavage:** Perfect **Fracture:** Splintery, conchoidal **Hardness:** 2 **Specific gravity:** 2.31 to 2.32 **Colour:** Colourless, white, grey, red, brown, yellowish, greenish **Streak:** White **Lustre:** Subvitreous, silky, pearly; transparent

▼ Gypsum from Yorkshire, England, UK.

Forming in evaporite deposits, where marine waters or inland lakes have dried out, gypsum also occurs around hot volcanic springs, in amygdales and as an alteration product in various geological environments. Satin spar is the name given to the form that has a fibrous habit, while transparent crystals are called selenite. A variety called desert rose consists of rosettes of crystals, often covered in a thin layer of sand; radiating crystal masses have the informal name daisy gypsum. Gypsum is found with many other evaporite minerals, especially halite and anhydrite. At the Naica mine at Chihuahua, Mexico, huge gypsum crystals up to 11 m (36 ft) long occur. Gypsum has many industrial uses, especially in the manufacture of plasterboard and plaster. It is soluble in hot water and hydrochloric acid.

▲ Daisy gypsum
from Stamp Hill Mine,
Cumbria, England, UK.

▶ Gypsum var.
desert-rose from
Ouargla Province,
Algeria.

Anhydrite

Chemical composition: $CaSO_4$ **Crystal system:** Orthorhombic **Mineral habit:** Crystals rare, tabular or prismatic; usually massive, granular or fibrous **Cleavage:** Perfect **Fracture:** Splintery, uneven **Hardness:** 3 to 3½ **Specific gravity:** 2.98 **Colour:** Colourless, white, grey, bluish, violet, pinkish, brown, reddish **Streak:** White **Lustre:** Vitreous, pearly, greasy; transparent to translucent

Anhydrite forms in evaporite sequences and less commonly in hydrothermal veins and in the cap rocks of salt domes. In evaporite deposits it is found with halite, sylvite, gypsum, dolomite and calcite. Anhydrite can occur in sequences of considerable thickness and is an economically important mineral, especially in the construction industry, where it is used in the manufacture of

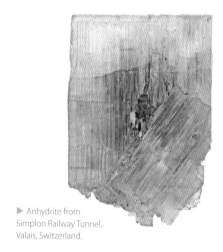

▶ Anhydrite from Simplon Railway Tunnel, Valais, Switzerland.

plaster and plasterboard. This mineral is closely related to gypsum but lacks water molecules in its chemistry. It melts easily, colouring a flame red.

Celestine

Chemical composition: $SrSO_4$ **Crystal system:** Orthorhombic **Mineral habit:** Crystals tabular, sometimes lath-like; also fibrous nodular, granular, earthy **Cleavage:** Perfect **Fracture:** Uneven **Hardness:** 3 to 3½ **Specific gravity:** 3.96 to 3.98 **Colour:** Colourless, white, grey, blue, green, red, brown, yellowish, orange **Streak:** White **Lustre:** Vitreous, pearly; transparent to translucent

Celestine (sometimes known as celestite) occurs in sedimentary rocks, especially limestones, in hydrothermal veins and sometimes in vesicles in lava. It is commonly found with halite, anhydrite and gypsum. Celestine is a source of strontium, which has a use in fireworks and flares, giving a brilliant red colour, and in the manufacture of specialized glass. This mineral is slightly soluble in acids and water and melts easily.

▶ Celestine from Pylle Hill, Gloucestershire, England, UK.

Barite

Chemical composition: $BaSO_4$ **Crystal system:** Orthorhombic **Mineral habit:** Crystals tabular, prismatic and in rosettes (desert roses) and fan-shaped crystal masses (cockscomb barite); also massive, granular, stalactitic, concretionary, earthy **Cleavage:** Perfect **Fracture:** Uneven **Hardness:** 3 **Specific gravity:** 4.50 **Colour:** Colourless, white, grey, bluish, greenish, reddish, brown, yellowish **Streak:** White **Lustre:** Vitreous to resinous or pearly; transparent to subtranslucent

Barite (alternatively spelled baryte or barytes) is a well-known and common mineral in hydrothermal veins, with fluorite, galena, dolomite, sphalerite, pyrite and quartz, where it is often considered a gangue mineral. It also forms in sedimentary rocks, in nodular masses, and around hot springs. Barite is an important ore of barium when found in large concentrations and has an economic use as a component of drilling mud in the oil and gas industry and in radiography. Barite melts with difficulty and does not dissolve in acids.

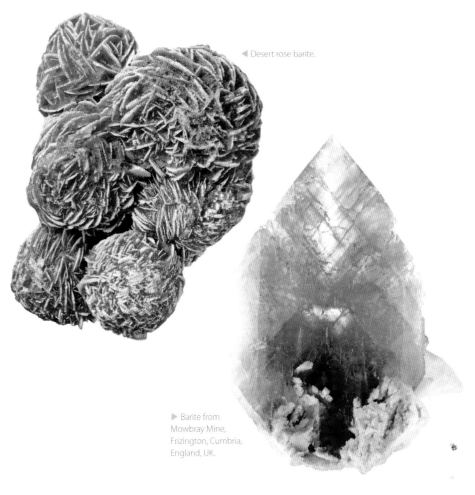

◀ Desert rose barite.

▶ Barite from Mowbray Mine, Frizington, Cumbria, England, UK.

Anglesite

Chemical composition: $PbSO_4$ **Crystal system:** Orthorhombic **Mineral habit:** Crystals tabular or prismatic; also massive, granular, stalactitic, nodular, encrusting, globular **Cleavage:** Good **Fracture:** Conchoidal **Hardness:** 2½ to 3 **Specific gravity:** 6.37 to 6.39 **Colour:** Colourless, white, grey, yellowish, pale blue, pale green **Streak:** Colourless **Lustre:** Adamantine, vitreous, resinous, transparent to opaque

Anglesite forms as a secondary mineral in veins containing lead, by the oxidation and alteration of minerals such as galena. In these environments it frequently occurs with sphalerite, chalcopyrite, calcite, barite, cerussite and quartz. It is sometimes found in layers and crusts around a core of galena. Anglesite may fluoresce a yellowish colour under ultraviolet light. It melts very easily when held in a flame, and dissolves slowly in nitric acid.

◀ Anglesite from Derbyshire, England, UK.

Chalcanthite

Chemical composition: $CuSO_4.5H_2O$ **Crystal system:** Triclinic **Mineral habit:** Crystals prismatic, tabular; also massive, fibrous, granular, encrusting, stalactitic **Cleavage:** Imperfect **Fracture:** Conchoidal **Hardness:** 2½ **Specific gravity:** 2.29 **Colour:** Sky-blue, dark blue, greenish-blue, green **Streak:** Colourless **Lustre:** Vitreous to resinous; transparent to translucent

Chalcanthite occurs where copper ores, especially sulphides such as chalcopyrite, have been altered by oxidation. This can be either by deep weathering or by the upward movement of hydrothermal fluids. It is associated with chalcopyrite, calcite, aragonite, malachite, azurite and brochantite. Chalcanthite can be found as an encrustation on the walls and roofs of mines, usually copper mines. It is soluble in water and will slowly degrade over time.

▼ Chalcanthite from Wheal Sperries, Cornwall, England, UK.

Melanterite

Chemical composition: $FeSO_4.7H_2O$ **Crystal system:** Monoclinic **Mineral habit:** Crystals rare, prismatic, tabular, octahedral; usually as fibrous efflorescences, stalactitic, encrusting **Cleavage:** Perfect **Fracture:** Conchoidal **Hardness:** 2 **Specific gravity:** 1.89 **Colour:** Green, blue, white **Streak:** Colourless **Lustre:** Vitreous or silky; translucent to opaque

◀ Melanterite from Vulcano, Lipari Islands, Italy.

Occurring as a secondary mineral, often in veins where minerals such as marcasite and pyrite have been altered by oxidation and circulating fluids, melanterite also forms as an efflorescence or coating on the walls of mine workings and rarely around volcanic fumaroles. It is found with minerals such as chalcanthite, epsomite and a variety of sulphate minerals, including copiapite. Melanterite is soluble in water and becomes magnetic when heated.

Jarosite

Chemical composition: $KFe^{3+}_3(SO_4)_2(OH)_6$ **Crystal system:** Trigonal **Mineral habit:** Crystals tiny, pseudocubic, tabular; also massive, fibrous, granular, earthy **Cleavage:** Distinct **Fracture:** Conchoidal to uneven **Hardness:** 2½ to 3½ **Specific gravity:** 2.90 to 3.26 **Colour:** Yellow, yellowish-brown, brown **Streak:** Pale yellow **Lustre:** Vitreous to resinous; translucent

Jarosite forms as a secondary mineral in iron-bearing deposits, where circulating fluids have altered primary minerals, and where weathering occurs, especially in arid climates. It is often found where pyrite decomposes. Jarosite also forms near hot springs and has been identified in acidic porphyritic rocks which have been altered by tourmalinization. This mineral is insoluble in water but soluble in hydrochloric acid. Jarosite fluoresces under ultraviolet light.

◀ Jarosite from Esperanza Mine, Jaroso Ravine, Sierra Almagrera, Almeria, Spain.

Glauberite

Chemical composition: $Na_2Ca(SO_4)_2$ **Crystal system:** Monoclinic **Mineral habit:** Crystals tabular, prismatic, dipyramidal, often striated; also as compact masses and crusts **Cleavage:** Perfect **Fracture:** Conchoidal **Hardness:** 2½ to 3 **Specific gravity:** 2.75 to 2.85 **Colour:** Colourless, yellowish, grey **Streak:** White **Lustre:** Vitreous, pearly; transparent to translucent

Glauberite occurs in various geological situations, including evaporite sequences, with halite, gypsum, anhydrite and polyhalite, and on the margins of salt lakes, nitrate deposits, fumaroles and in vesicles in basalts. It will dissolve in hydrochloric acid, and melts easily when held in a flame.

▶ Glauberite from Borax Lake, California, USA.

Thenardite

Chemical composition: Na_2SO_4 **Crystal system:** Orthorhombic **Mineral habit:** Crystals tabular, dipyramidal, rarely prismatic, often twinned; also as crusts and efflorescences **Cleavage:** Perfect **Fracture:** Uneven, hackly **Hardness:** 2½ to 3 **Specific gravity:** 2.66 **Colour:** Colourless, greyish, brownish, reddish, yellowish **Streak:** White **Lustre:** Vitreous to resinous; transparent to translucent

Forming in playas and salt-lake deposits, associated with gypsum, halite, glauberite and epsomite, Thenardite also occurs as an efflorescence on the soil surface in arid regions and can be found around fumaroles and associated with lavas. This mineral is easily dissolved in water and has a salty taste.

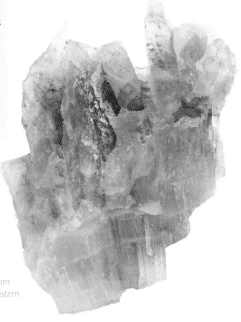

▶ Thenardite from Wadi Natrun, Western Desert, Egypt.

Polyhalite

Chemical composition: $K_2MgCa_2(SO_4)_4 \cdot 2H_2O$
Crystal system: Triclinic **Mineral habit:**
Crystals rare, tabular, small, commonly twinned;
usually massive, foliated, fibrous **Cleavage:**
Perfect **Fracture:** Uneven, hackly **Hardness:**
2½ to 3½ **Specific gravity:** 2.78 **Colour:**
Colourless, white, grey; often brownish,
reddish or pink because of iron inclusions,
yellowish **Streak:** White **Lustre:** Vitreous to
resinous; transparent to translucent

Polyhalite is found in evaporite deposits
in sedimentary rocks, with halite, sylvite,
gypsum, anhydrite, carnallite and other
evaporite minerals. Rarely, it forms around
volcanic vents. It has a rather sharp,

▼ Polyhalite.

bitter taste. Polyhalite is of considerable
economic importance, because it is used to
manufacture fertilizers.

Linarite

Chemical composition: $PbCu(SO_4)(OH)_2$
Crystal system: Monoclinic **Mineral habit:**
Crystals tabular or prismatic, often twinned;
also as crusts and aggregates **Cleavage:**
Perfect **Fracture:** Conchoidal **Hardness:** 2½
Specific gravity: 5.35 **Colour:** Dark blue
Streak: Pale blue **Lustre:** Vitreous to
subadamantine; transparent to translucent

Linarite forms where veins and other deposits
containing lead and copper have been oxidized
and altered. It occurs with a range of minerals
including chalcanthite, brochantite and anglesite.
Its striking colouring can cause it to be confused
with azurite, but linarite is less hard at 2½,
compared with azurite at 3½ to 4, and linarite has
a higher specific gravity, 5.35 compared to 3.77.
Linarite will dissolve in nitric acid, and melts when
held in a flame, turning black.

▶ Linarite from
Roughten Gill,
Cumbria, England, UK.

Caledonite

Chemical composition: $Cu_2Pb_5(SO_4)_3CO_3(OH)_6$
Crystal system: Orthorhombic **Mineral habit:** Crystals small, prismatic, often striated; also massive and as coatings **Cleavage:** Perfect **Fracture:** Uneven **Hardness:** 2½ to 3 **Specific gravity:** 5.75 to 5.77 **Colour:** Pale blue-green, green-blue **Streak:** Greenish blue to bluish white **Lustre:** Vitreous to resinous; transparent to translucent

Caledonite occurs with many other secondary minerals in the oxidized parts of lead and copper ore deposits. It is found with linarite, cerussite, leadhillite, anglesite, brochantite, malachite, azurite and quartz. Caledonite is a relatively rare mineral that is valued by collectors for its colour.

▶ Caledonite from Leadhills, South Lanarkshire, Scotland, UK.

Copiapite

Chemical composition:
$Fe^{2+}Fe^{3+}_4(SO_4)_6(OH)_2 \cdot 20H_2O$ **Crystal system:**
Triclinic **Mineral habit:** Crystals tabular; usually
as scaly aggregates, also as crusts and grains
Cleavage: Perfect **Fracture:** Uneven
Hardness: 2½ to 3 **Specific gravity:** 2.08 to 2.17
Colour: Yellow, golden yellow, orange yellow;
greenish yellow, olive green **Streak:** Pale yellow
Lustre: Pearly; transparent to translucent

◀ Copiapite.

Copiapite usually forms by the alteration of
sulphides such as pyrite, commonly where
oxidation takes place in arid climates. It often
occurs with melanterite and other water-rich
sulphates. Copiapite is soluble in water and melts
at a relatively low temperature.

Cyanotrichite

Chemical composition: $Cu_4Al_2(SO_4)(OH)_{12} \cdot 2H_2O$
Crystal system: Monoclinic **Mineral habit:**
Crystals very small, acicular, forming coatings
and aggregates; also as tufts and veinlets
Cleavage: Good **Fracture:** Uneven **Hardness:**
1 to 3 **Specific gravity:** 2.76 **Colour:** Pale sky-
blue to azure-blue **Streak:** Pale blue **Lustre:**
Silky; translucent or transparent

Cyanotrichite is a rare mineral that occurs where
copper ore deposits have been weathered
and oxidized. It is associated with brochantite,
chalcophyllite, azurite, malachite and olivenite.
Cyanotrichite is valued by collectors for its
extremely slender, fibrous crystals, which can
encrust the whole surface of a specimen, giving
it a velvet-like appearance. It dissolves in acids
and melts in a flame.

▶ Cyanotrichite from
Grand View Mine,
Grand Canyon,
Coconino County,
Arizona, USA.

Brochantite

Chemical composition: $Cu_4(SO_4)(OH)_6$ **Crystal system:** Monoclinic **Mineral habit:** Crystals prismatic, acicular, tabular; also encrusting, granular, massive **Cleavage:** Perfect **Fracture:** Conchoidal to uneven **Hardness:** 3½ to 4 **Specific gravity:** 3.97 **Colour:** Bright green to blackish-green, pale green **Streak:** Pale green **Lustre:** Vitreous or pearly; transparent to translucent

Brochantite commonly forms where deposits containing copper minerals have been altered by weathering and oxidation. Although widespread, larger concentrations are found in arid conditions. Brochantite occurs with malachite and azurite. This mineral is soluble in both hydrochloric and nitric acids, and melts when held in a flame.

▲ Brochantite from Wales, UK.

▶ Brochantite from Milpillas Mine, Sonora, Mexico.

Ettringite

Chemical composition: $Ca_6Al_2(SO_4)_3(OH)_{12}.26H_2O$ **Crystal system:** Trigonal **Mineral habit:** Crystals, prismatic, rhombohedral, dipyramidal; also fibrous **Cleavage:** Perfect **Fracture:** Uneven **Hardness:** 2 to 2½ **Specific gravity:** 1.77 **Colour:** Colourless, white, pale yellow **Streak:** White **Lustre:** Vitreous; transparent

Ettringite occurs in cavities in limestones where these have been enclosed in lava and metamorphosed. In such situations very well-formed crystals can be found. It is a member of the ettringite group of minerals, which comprises a small number of calcium sulphate minerals containing aluminium, chromium, iron, manganese and silicon. For example, bentorite contains chromium instead of aluminium, while sturmanite has iron, manganese and boron in its chemistry.

▶ Ettringite from N'Chwaning Mines, Kuruman, Northern Cape, South Africa.

Chromates, molybdates and tungstates

Crocoite

Chemical composition: PbCrO$_4$ **Crystal system:** Monoclinic **Mineral habit:** Crystals prismatic, octahedral or rhombohedral, often as aggregates of slender crystals referred to as 'jackstraw' masses; also massive **Cleavage:** Distinct **Fracture:** Conchoidal to uneven **Hardness:** 2½ to 3 **Specific gravity:** 5.97 to 6.02 **Colour:** Red-orange, orange, red, yellow **Streak:** Orange-yellow **Lustre:** Adamantine to vitreous; transparent to translucent

This chromate mineral forms where deposits containing lead and chromium have been altered by weathering and oxidation. Crocoite occurs with many other secondary minerals, including cerussite, vanadinite, wulfenite and pyromorphite. It melts in a flame and is soluble only in strong acids. Because of its bright colouring and its frequent occurrence as superb crystals, crocoite is prized by collectors and was once used as a pigment.

▲ Crocoite from Berezovsk Deposit, Sverdlovsk Oblast, Russia.

Wulfenite

Chemical composition: PbMoO$_4$ **Crystal system:** Tetragonal **Mineral habit:** Crystals tabular, often with a square outline, octahedral, prismatic; also massive, granular **Cleavage:** Distinct **Fracture:** Uneven to subconchoidal **Hardness:** 2½ to 3 **Specific gravity:** 6.50 to 7.50 **Colour:** Yellow, orange, brown, yellowish-grey, pink, greenish-brown, blue **Streak:** White **Lustre:** Resinous to adamantine; transparent to translucent

▼ Wulfenite from Helena Mine, Mežica, Carinthia, Slovenia.

Wulfenite is a molybdate that forms as a secondary mineral where ore deposits, especially those containing lead, have been altered by oxidation. It commonly occurs with galena, pyromorphite, cerussite, calcite, limonite, malachite, mimetite and vanadinite. Wulfenite is soluble in heated hydrochloric acid and melts easily in a flame.

Scheelite

Chemical composition: Ca(WO$_4$) **Crystal system:** Tetragonal **Mineral habit:** Crystals octahedral, tabular, often striated and twinned; also massive, granular, columnar **Cleavage:** Distinct **Fracture:** Subconchoidal to uneven **Hardness:** 4½ to 5 **Specific gravity:** 6.10 **Colour:** Colourless, white, grey, yellowish, brownish, greenish, orange-yellow, reddish **Streak:** White **Lustre:** Vitreous to adamantine; transparent to translucent

◀ Scheelite from Cumbria, England, UK.

Scheelite is a tungstate belonging to the scheelite group of minerals, which includes powellite, wulfenite and stolzite. These minerals contain calcium, lead, molybdenum and tungsten. Scheelite occurs in hydrothermal veins and pegmatites and in rocks altered by contact metamorphism. It is also found in placer sands. This mineral will fluoresce white or bluish-white under ultraviolet light. It melts only with difficulty, but is soluble in acids. Scheelite is an important ore of tungsten. Powellite is the molybdenum-dominant equivalent of scheelite, being Ca(MoO$_4$). It is softer than scheelite at 3½ to 4 on the Mohs scale, and has a lower specific gravity (4.26). It fluoresces yellow in ultraviolet light. The tungstate mineral stolzite has a hardness of 2½ to 3 and a high specific gravity of 8.34. It is the lead-dominant equivalent of scheelite with the chemical formula Pb(Wo$_4$).

Phosphates, arsenates and vanadates

Phosphates are formed when metals combine with the phosphate anion (PO_4). Many phosphates are secondary minerals, being the result of the oxidation of sulphides. Arsenates are produced by the combination of metals with the arsenate anion (AsO_4). Many arsenates are brightly coloured and exhibit fine crystals, making them prized by collectors. Vanadates form when metals combine with the vanadate anion (VO_4).

Phosphates

Amblygonite

Chemical composition: $LiAl(PO_4)F$ **Crystal system:** Triclinic **Mineral habit:** Crystals prismatic, commonly twinned; also massive **Cleavage:** Perfect **Fracture:** Conchoidal to uneven **Hardness:** 5½ to 6 **Specific gravity:** 3.04 to 3.11 **Colour:** White, greyish-white, colourless, yellowish, pinkish, greenish, bluish **Streak:** White **Lustre:** Vitreous to greasy or pearly; transparent to translucent

This mineral is found in granitic pegmatites, sometimes as very large crystals weighing several tonnes. Amblygonite also occurs as large masses, which can be up to 7 m (23 ft) long and 3 m (10 ft) wide, and weigh up to 200 tonnes. It melts easily and colours the flame red because of its lithium content.

▶ Amblygonite from Taquaral, Itinga, Minas Gerais, Brazil.

Pyromorphite

Chemical composition: $Pb_5(PO_4)_3Cl$ **Crystal system:** Hexagonal **Mineral habit:** Crystals prismatic, hexagonal, barrel-shaped, sometimes tabular, pyramidal; also globular, reniform, botryoidal, granular, earthy **Cleavage:** Very poor **Fracture:** Uneven to subconchoidal **Hardness:** 3½ to 4 **Specific gravity:** 7.04 **Colour:** Green, yellow, grey, orange, brown, white, rarely red **Streak:** White **Lustre:** Subadamantine to resinous; transparent to translucent

▼ Pyromorphite from Bad Ems, Rhineland-Palatinate, Germany.

Pyromorphite forms in lead veins and other lead deposits, when primary lead-bearing minerals are altered by oxidation. It is thus a secondary mineral, and because of its fine crystals and colour, it is sought-after by collectors. Pyromorphite is soluble in acids and melts easily. It is an end member of the pyromorphite-mimetite series. Mimetite has a similar chemistry, but contains arsenic, rather than phosphorus, having a composition of $Pb_5(AsO_4)_3Cl$. Campylite is an unusual form of mimetite with barrel-shaped crystals. Although many pyromorphites are green and mimetites often are yellowish orange, it is impossible visually to tell the difference between the two minerals.

Lazulite

Chemical composition: $MgAl_2(PO_4)_2(OH)_2$ **Crystal system:** Monoclinic **Mineral habit:** Crystals pyramidal, tabular, often twinned; also massive, granular, compact **Cleavage:** Indistinct to good **Fracture:** Uneven **Hardness:** 5½ to 6 **Specific gravity:** 3.12 to 3.24 **Colour:** Deep azure-blue, pale blue, bluish-green **Streak:** White **Lustre:** Vitreous to dull; translucent to opaque, rarely transparent

◀ Lazulite.

Lazulite occurs in a number of geological environments, especially in quartz veins, pegmatites and quartzites. It is often found with quartz, kyanite, rutile, garnet, muscovite, corundum, andalusite, pyrophyllite and sillimanite. When heated, it breaks into small pieces, but does not melt. Very occasionally lazulite is used as a gemstone.

Vivianite

Chemical composition: $Fe^{2+}_3(PO_4)_2.8H_2O$
Crystal system: Monoclinic **Mineral habit:**
Crystals prismatic or tabular, often in clusters
or stellate groups; also massive, fibrous, bladed,
encrusting, concretionary, earthy **Cleavage:**
Perfect **Fracture:** Uneven **Hardness:** 1½
to 2 **Specific gravity:** 2.67 to 2.69 **Colour:**
Colourless, blue, green; darkens on exposure
to dark green, purplish, bluish-black **Streak:**
Colourless, changing to dark blue or brown
on exposure **Lustre:** Vitreous to pearly;
transparent to translucent

▼ Vivianite from
Trepča Stan Terg
Mine, Kosovska
Mitrovica, Kosovo.

Vivianite is a secondary mineral found in ore
deposits where primary minerals have been
oxidized. It also occurs in pegmatites where
phosphate minerals have been altered. In
sedimentary rocks, vivianite has been found
on fossilized bone and in the shells of fossil
molluscs. It often develops as fine crystals, but will
decompose unless stored correctly. Vivianite melts
easily in a flame and dissolves in strong acids.

Torbernite

▼ Torbernite (uranium
ore) from Margabal
Mine, France.

Chemical composition: $Cu(UO_2)_2(PO_4)_2.12H_2O$
Crystal system: Tetragonal **Mineral habit:**
Tabular, octagonal or rectangular, rarely
pyramidal; also as lamellar masses, granular,
earthy **Cleavage:** Perfect **Fracture:** Uneven
Hardness: 2 to 2½ **Specific gravity:** 3.22
Colour: Various shades of green **Streak:** Pale
green **Lustre:** Vitreous to subadamantine,
pearly; transparent to translucent

Torbernite forms in pegmatites and can occur as a
secondary mineral resulting from the alteration of
uraninite. Being radioactive, its uranium content
decays, and torbernite's chemical instability leads
to the development over time of a different
mineral called metatorbernite. Torbernite is
soluble in strong acids. It has been used as a
source of uranium, and appropriate care must be
taken when handling specimens.

Autunite

Chemical composition:
$Ca(UO_2)_2(PO_4)_2 \cdot 10\text{-}12H_2O$ **Crystal system:**
Orthorhombic **Mineral habit:** Crystals
tabular, octagonal or rectangular, often in
fan-shaped aggregates; also encrusting,
granular, earthy **Cleavage:** Perfect **Fracture:**
Uneven **Hardness:** 2 to 2½ **Specific gravity:**
3.05 to 3.20 **Colour:** Shades of yellow,
greenish yellow, pale to dark green **Streak:**
Pale yellow **Lustre:** Vitreous, pearly, dull;
transparent to translucent

This radioactive mineral often forms by the
secondary alteration of uraninite and is found in
pegmatites, hydrothermal veins and granites.
Autunite is a very important ore of uranium. In
ultraviolet light it fluoresces strongly, producing
a yellowish green colour. When heated, it alters to
meta-autunite. Extreme care must be taken when
handling radioactive specimens.

▶ Autunite from
Senhore De
Assuncao Mine,
Viseu, Portugal.

Xenotime-(Y)

Chemical composition: $Y(PO_4)$ **Crystal
system:** Tetragonal **Mineral habit:** Crystals
prismatic, pyramidal, as rosettes and
aggregates **Cleavage:** Perfect **Fracture:**
Splintery or uneven **Hardness:** 4 to 5 **Specific
gravity:** 4.40 to 5.10 **Colour:** Yellowish-brown,
reddish-brown; also pale yellow, pale grey,
reddish, greenish **Streak:** Pale brown **Lustre:**
Vitreous to resinous; translucent to opaque

▼ Xenotime crystals
in host rock found in
Tvedestrand, Norway.

Xenotime-(Y) forms in pegmatites and other
igneous rocks, especially granite and syenite.
It also occurs in high-grade metamorphic rocks,
with micas and feldspars. Other minerals with
which it is found include quartz, zircon, anatase,
rutile, siderite and apatite. Xenotime-(Y) can
also occur in quartz veins and as a detrital
mineral in sediments.

Monazite-(Ce)

Chemical composition: Ce(PO$_4$) **Crystal system:** Monoclinic **Mineral habit:** Crystals tabular, prismatic, often rough, striated, twinned and at times very large; also massive, granular **Cleavage:** Distinct **Fracture:** Conchoidal to uneven **Hardness:** 5 to 5½ **Specific gravity:** 5.00 to 5.50 **Colour:** Reddish-brown, brown, greyish-white, yellow, greenish, pink **Streak:** White **Lustre:** Resinous, waxy, vitreous to subadamantine; transparent to subtranslucent

Monazite-(Ce) belongs to the monazite group, which contains phosphate, arsenate and silicate minerals, including four types of monazite. The most common contains cerium (Ce). Other forms are monazite-(La) (lanthanum), monazite-(Nd)

▼ Monazite.

(neodymium) and monazite-(Sm) (samarium). Monazite occurs in pegmatites, high-grade metamorphic rocks and quartz veins. It is found in placer sands, where these accumulate on beaches and riverbeds. Monazite-(Ce) is the main ore of cerium and often contains traces of useful thorium. It is insoluble and does not melt.

Wavellite

Chemical composition: Al$_3$(PO$_4$)$_2$(OH)$_3$5H$_2$O **Crystal system:** Orthorhombic **Mineral habit:** Crystals prismatic, very small, rare, commonly as radiating masses, often spherical; also stalactitic and encrusting **Cleavage:** Perfect **Fracture:** Subconchoidal to uneven **Hardness:** 3½ to 4 **Specific gravity:** 2.36 **Colour:** White, greenish-white, green, yellow, yellowish-brown, brown, brownish-black, blue **Streak:** White **Lustre:** Vitreous to resinous or pearly; transparent to translucent

Wavellite forms as a secondary mineral in hydrothermal veins and some metamorphic rocks, especially as rosettes and radiating masses on the surfaces of rock fractures. Small globular structures are sometimes found, which, when cut open, reveal a fine radiating structure. For this reason the mineral is valued by collectors. Wavellite is soluble in strong acids, but cannot be melted.

▼ Wavellite from High Down Quarry, West Buckland, Devon, England, UK.

Turquoise

Chemical composition: $CuAl_6(PO_4)_4(OH)_8.4H_2O$
Crystal system: Triclinic **Mineral habit:**
Crystals small, prismatic, uncommon; usually
massive, granular, concretionary, stalactitic,
as crusts **Cleavage:** Perfect **Fracture:**
Conchoidal **Hardness:** 5 to 6 **Specific gravity:**
2.60 to 2.80 **Colour:** Bright blue, bluish-green,
apple-green, greenish-grey **Streak:** White
to pale green **Lustre:** Vitreous, waxy or dull;
transparent to opaque

Turquoise is a secondary mineral formed when
rocks rich in aluminium-bearing minerals are
altered, usually by waters from the surface seeping
into them, especially during weathering and
when copper is also present. This mineral has
been used for many years as a gemstone, primarily
because of its blue colouring. It is relatively soft
for this purpose, but it takes on a good lustre
when polished. Some specimens have the added
attraction of containing small specks of pyrite.
Recently, its use as a gemstone has been overtaken
by synthetic imitations. Turquoise is soluble in
heated hydrochloric acid, but cannot be melted.

◄ Turquoise from
Wadi Maghara, South
Sinai, Egypt.

► Turquoise from
Lynch Station,
Campbell County,
Virginia, USA.

Apatite group

Chemical composition: $Ca_5(PO_4)_3(F,Cl,OH)$
Crystal system: Hexagonal   **Mineral habit:**
Crystals prismatic, tabular; also massive,
granular, as botryoidal crusts   **Cleavage:** Poor
Fracture: Conchoidal to uneven   **Hardness:** 5
Specific gravity: 3.16 to 3.22   **Colour:** Greenish,
yellow, brown, white, colourless, bluish, purple,
pink   **Streak:** White   **Lustre:** Vitreous to
subresinous; transparent to opaque

Apatite is strictly the name given to a complex
supergroup of minerals and a smaller group of
calcium phosphates. It is impossible visually to
determine which specific member of the apatite
supergroup is present, so geologists tend to call
them all apatite unless they have been definitively
tested. Apatite defines point five on the Mohs
scale of hardness. The supergroup includes
not only phosphates, but also arsenates and
vanadates. They contain a variety of chemical
elements, including barium, cerium, strontium
and yttrium. Apatite supergroup minerals form
in igneous and metamorphic rocks, especially
marble. Their main use is in the production
of phosphate fertilizers. The specific mineral
hydroxyapatite is the same material that makes
up our bones and teeth. Some apatite minerals
display yellow fluorescence under ultraviolet light.

▶ Apatite from
Durango, Mexico.

▶ Apatite from
Mount Apatite,
Maine, USA.

Variscite

Chemical composition: $Al(PO_4).2H_2O$ **Crystal system:** Orthorhombic **Mineral habit:** Crystals octahedral, rare; usually massive, as crusts or nodules **Cleavage:** Perfect **Fracture:** Conchoidal, uneven, splintery **Hardness:** 3½ to 4½ **Specific gravity:** 2.56 to 2.61 **Colour:** Pale green, emerald-green, bluish-green, colourless, rarely pink **Streak:** White **Lustre:** Vitreous, waxy, dull; transparent to translucent

Variscite is a widespread mineral, which forms when phosphate-rich waters react with rocks containing aluminium. Though it can have a similar appearance to turquoise, variscite is usually greener in colour. It is sometimes used as a gemstone. Variscite cannot be melted, but will dissolve in acids if heated.

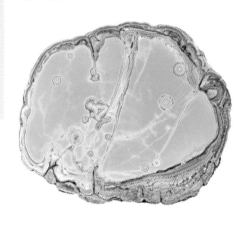

▼ Variscite - polished surface of cut nodule - from Fairfield, Utah, USA.

Brazilianite

Chemical composition: $NaAl_3(PO_4)_2(OH)_4$ **Crystal system:** Monoclinic **Mineral habit:** Crystals prismatic, can be spear-shaped; also globular with internal radiating fibrous structure **Cleavage:** Good **Fracture:** Conchoidal **Hardness:** 5½ **Specific gravity:** 2.98 **Colour:** Colourless, pale yellowish-green, pale yellow **Streak:** Colourless **Lustre:** Vitreous; transparent

Brazilianite forms in pegmatites where hydrothermal fluids have risen into cavities. It occurs with beryl, quartz, feldspars, apatite, tourmaline and micas. Brazilianite was discovered relatively recently, originally described about 70 years ago, and at first confused with chrysoberyl. It has been used as a gemstone, but is mainly of interest to mineral collectors. Brazilianite dissolves with difficulty in strong acids.

▼ Brazilianite from Minas Gerais, Brazil.

Arsenates

Adamite

Chemical composition: $Zn_2(AsO_4)(OH)$　**Crystal system:** Orthorhombic　**Mineral habit:** Crystals tabular, often with wedge-shaped terminations; also as interlocking, radial or spherical aggregates　**Cleavage:** Good　**Fracture:** Subconchoidal to uneven　**Hardness:** 3½　**Specific gravity:** 4.32 to 4.48　**Colour:** Yellow-green, greenish, yellowish, pinkish, white, colourless, bluish　**Streak:** White　**Lustre:** Vitreous; transparent to translucent

▼ Adamite from Ojuela Mine, Mapimí, Durango, Mexico.

Adamite is a secondary mineral that forms by the alteration of ore deposits, especially veins containing zinc minerals. It is often found with malachite, azurite, calcite, hemimorphite and smithsonite. Adamite melts in a flame, dissolves in dilute acids and sometimes fluoresces yellow in ultraviolet light.

Annabergite

Chemical composition: $Ni_3(AsO_4)_2.8H_2O$　**Crystal system:** Monoclinic　**Mineral habit:** Crystals prismatic, often with striated faces; usually as crystalline crusts and earthy or powdery masses　**Cleavage:** Perfect　**Fracture:** Uneven　**Hardness:** 1½ to 2½　**Specific gravity:** 3.07　**Colour:** White, greyish, pale green, yellowish-green, pinkish, brownish　**Streak:** Paler than colour of mineral　**Lustre:** Adamantine, pearly or dull; transparent to translucent

▶ Annabergite from Lavrion Mining District, Attica, Greece.

This secondary mineral forms where veins containing nickel have been oxidized. Annabergite is closely related to erythrite, which has a very similar chemistry to annabergite, but contains cobalt rather than nickel, with the composition $Co_3(AsO_4)_2.8H_2O$. One of the main identification differences between the two minerals is their colour: erythrite is purple red to deep pink, compared with annabergite's greenish and yellowish colours, although only a little cobalt within an annabergite specimen may change its colour to slightly pinkish. The hardness is the same.

▼ Erythrite from Queensland, Australia.

The specific gravity of erythrite is 3.06. Both these minerals are soluble in acids and melt easily in a flame.

Clinoclase

Chemical composition: $Cu_3(AsO_4)(OH)_3$
Crystal system: Monoclinic **Mineral habit:**
Crystals tabular, elongated, rhombohedral; also
as rosettes with a rounded outline and radiating
structure **Cleavage:** Perfect **Fracture:**
Uneven **Hardness:** 2½ to 3 **Specific gravity:**
4.38 **Colour:** Dark bluish to black **Streak:**
Greenish-blue **Lustre:** Vitreous, pearly;
transparent to translucent

▼ Clinoclase from
St. Day, Cornwall,
England, UK.

Clinoclase is a secondary mineral that occurs
where copper deposits have been altered by
weathering and oxidation, clinoclase is an
uncommon mineral, often found with olivenite, a
mineral with a similar chemical composition. As a
result of its arsenic content, clinoclase gives off a
smell of garlic when heated. It is soluble in acids.

Olivenite

Chemical composition: $Cu_2(AsO_4)(OH)$
Crystal system: Monoclinic **Mineral habit:**
Crystals prismatic, acicular, tabular; also
reniform, globular, massive, granular, earthy,
nodular **Cleavage:** Indistinct **Fracture:**
Uneven to conchoidal **Hardness:** 3 **Specific
gravity:** 4.46 **Colour:** Olive-green, brownish-
green, brown, yellowish, grey-green, white
Streak: Yellowish-green **Lustre:** Vitreous to
adamantine, pearly, silky; translucent to opaque

▼ Olivenite from
Wheal Gorland,
St. Day, Cornwall,
England, UK.

Olivenite is a secondary mineral that forms where
copper ore veins have been altered, occurring
with a wide variety of other minerals including
malachite, azurite, goethite, calcite and dioptase.
It is common as a crust, but quite rare as well-
formed crystals and acicular aggregates. Like
clinoclase, a chemically similar mineral, olivenite
melts in a flame, producing a smell of garlic, and it
dissolves in acids.

Mimetite

Chemical composition: $Pb_5(AsO_4)_3Cl$
Crystal system: Hexagonal **Mineral habit:**
Crystals prismatic, acicular; usually globular,
botryoidal, reniform, granular **Cleavage:**
None **Fracture:** Subconchoidal to
uneven **Hardness:** 3½ to 4 **Specific gravity:**
7.24 **Colour:** Yellow, brownish-yellow, orange-
yellow, orange, red, white, colourless **Streak:**
White **Lustre:** Subadamantine to resinous;
transparent to translucent

This secondary mineral forms where veins and
other deposits containing lead have been altered
by oxidation. Mimetite is found with a variety of
other minerals, including galena, pyromorphite,
quartz, arsenopyrite, hemimorphite and
vanadinite. A variety of mimetite called campylite

is known for occurring as barrel-shaped crystals.
Mimetite is closely related to pyromorphite, a
phosphate of lead. Mimetite dissolves in acids
and melts easily, producing a smell of garlic
because of its arsenic content.

▲ Mimetite var.
Campylite from Dry
Gill Mine, Caldbeck,
Cumbria, England, UK.

▶ Mimetite from
Johanngeorgenstadt,
Saxony, Germany.

Scorodite

Chemical composition: $Fe^{3+}(AsO_4).2H_2O$
Crystal system: Orthorhombic **Mineral habit:**
Crystals pyramidal, tabular, prismatic:
also massive, earthy, as crusts **Cleavage:**
Imperfect **Fracture:** Subconchoidal
Hardness: 3½ to 4 **Specific gravity:** 3.27
Colour: Grey-green, yellowish-brown, brown,
bluish-green, blue, violet, colourless **Streak:**
Greenish-white **Lustre:** Vitreous, resinous or
dull; transparent to translucent

▼ Scorodite on Quartz
from Hemerdon Mine,
Devon, England, UK.

Scorodite occurs as a common secondary mineral
where other arsenic-rich minerals have been
altered by oxidation. It can be found where
cracked and crumbling arsenopyrite occurs, and in
this situation forms crystalline crusts. It is soluble
in hydrochloric and nitric acids. When scorodite is
weathered, limonite may form.

Bayldonite

Chemical composition: $Cu_3PbO(AsO_3OH)_2(OH_2)$
Crystal system: Monoclinic **Mineral habit:**
Rarely crystalline; commonly massive, powdery
or fine grained; also as crusts and concretions
with a fibrous structure **Cleavage:** None
Fracture: Uneven **Hardness:** 4½ **Specific
gravity:** 5.24 to 5.65 **Colour:** Bright grass-
green, yellowish-green **Streak:** Green
Lustre: Resinous; subtranslucent

▼ Bayldonite on
Quartz from United
Mines, St Day, Cornwall,
England, UK.

Bayldonite is an unusual secondary mineral,
which forms when copper ore deposits are
altered by oxidation in the presence of
lead and arsenic. When this happens,
it occurs with a number of minerals
including azurite, malachite, anglesite,
barite, cerussite, wulfenite, olivenite
and mimetite. Bayldonite gives off water
when heated in a closed test tube.

Pharmacosiderite

Chemical composition: $KFe^{3+}_4(AsO_4)_3(OH)_4 \cdot 6-7H_2O$ **Crystal system:** Cubic **Mineral habit:** Crystals cubic, often with diagonally striated faces, tetrahedral; also granular, earthy **Cleavage:** Imperfect **Fracture:** Uneven **Hardness:** 2½ **Specific gravity:** 2.80 **Colour:** Emerald green, olive green, yellow, reddish-brown, dark brown **Streak:** Yellowish-green **Lustre:** Adamantine to greasy; transparent to translucent

▼ Pharmacosiderite from Wheal Gorland, St. Day, Cornwall, England, UK.

Pharmacosiderite is a relatively rare mineral that forms where deposits containing arsenic have been altered by oxidation. This typically occurs where minerals such as arsenopyrite and tennantite have been altered. It can also be found in hydrothermal veins. Associated minerals include scorodite and beudantite.

Vanadates

Vanadinite

Chemical composition: $Pb_5(VO_4)_3Cl$ **Crystal system:** Hexagonal **Mineral habit:** Crystals prismatic, acicular, sometimes hollow or hopper-shaped; also globular **Cleavage:** None **Fracture:** Conchoidal to uneven **Hardness:** 2½ to 3 **Specific gravity:** 6.88 **Colour:** Bright red, orange-red, brownish-red, pale yellow, brownish-yellow **Streak:** White or yellowish **Lustre:** Resinous to subadamantine; transparent to opaque

Vanadinite forms when veins and other deposits containing lead are altered by oxidation. It is a potential source of vanadium, which is used in industry to produce steel alloys, but most vanadium is obtained from other minerals. Vanadinite is prized by mineral collectors because of its striking colour and crystalline habit. It melts easily in a flame and dissolves in nitric acid, leaving a reddish residue after the liquid has evaporated.

▼ Vanadinite from Mibladen, Midelt, Morocco.

Descloizite

Chemical composition: PbZn(VO$_4$)(OH)
Crystal system: Orthorhombic **Mineral habit:**
Crystals commonly pyramidal, prismatic or
tabular, often with uneven or rough surfaces;
also stalactitic, botryoidal, massive, granular,
as crusts **Cleavage:** None **Fracture:** Uneven
to conchoidal **Hardness:** 3 to 3½ **Specific
gravity:** 6.20 **Colour:** Orange-red, dark
red-brown, blackish-brown, dark green,
black **Streak:** Yellowish-orange to reddish-
brown **Lustre:** Vitreous to greasy; transparent
to opaque

Descloizite gives its name to a small group of
arsenate and vanadate minerals that include
mottramite. This has a very similar chemical
composition to descloizite, but is copper-dominant
rather than zinc-dominant, being PbCu(VO$_4$)
(OH). It has a lower specific gravity at 5.90. Both
crystallize in the orthorhombic system and occur as
pyramidal, prismatic or tabular crystals. Descloizite
and mottramite form as secondary minerals
where ore deposits have been altered. They
are found with many other secondary minerals
including vanadinite, pyromorphite, mimetite,
calcite, and cerussite. Descloizite is soluble in acids
and melts easily in a flame.

▼ Descloizite
from Berg Aukas
Mine, Grootfontein,
Namibia.

Silicates

Silicate minerals constitute the largest and most abundant class. Many silicates are very important as rock-formers. Igneous rocks such as granite and basalt, for example, are almost completely composed of silicate minerals plus quartz (silicon dioxide). Because silicate minerals generally have a high hardness, they resist erosion and become incorporated into sedimentary rocks. Many metamorphic rocks also contain primary silicate minerals. Chemically, silicates are compounds of metals in a variety of structural arrangements combined with the silicon-oxygen tetrahedron, SiO_4. These minerals can have very complex formulae.

Minerals in the pyroxene group are silicates with very similar internal structures, containing metallic elements, especially iron, magnesium and calcium, and lacking hydroxide. They are important as rock-forming minerals in basic and ultrabasic igneous rocks and also in some metamorphic rocks. Crystals of the subgroup orthopyroxenes are classified in the orthorhombic system, while those of the subgroup clinopyroxenes are monoclinic. Pyroxenes usually have two good cleavage planes which intersect at nearly 90°.

The minerals of the amphibole supergroup can have very complex formulae but generally are hydrous silicates of metallic elements. They occur as rock-formers in intermediate and acid igneous rocks and are also found in many metamorphic rocks. As with the pyroxene group, there are two good cleavage surfaces, but in the amphibole minerals these have an angle of about 120° between them.

Mica group minerals are common in acid igneous rocks, metamorphic rocks and some sedimentary rocks. The members of the group are often clearly made up of thin sheets that break apart easily, making them very flaky. In the study of rocks, it is useful to distinguish light- and dark-coloured micas, this difference being a reflection of their chemistry; the lighter ones

contain aluminium and sometimes lithium, and the darker ones contain magnesium and iron.

Feldspars comprise the most abundant group of minerals, occurring as essential, primary components in many igneous and metamorphic rocks. Two main groups are recognized, the plagioclase and potassic feldspars. The plagioclase feldspars are aluminium silicates of sodium and calcium, and these elements can substitute for each other, producing a series of minerals. This substitution of one element for another is referred to as solid solution. Albite is the sodium-rich end-member, while the calcium-rich end-member is anorthite. There are a number of named varieties between these end-members; labradorite, for example, is a silicate of both sodium and calcium.

Potassic feldspars are aluminium silicates of potassium. Orthoclase, sanidine and microcline are three common minerals in this group. These feldspars tend to form as primary components in acid igneous rocks, such as rhyolites and granites. They also occur as schists and gneisses and as grains in some sedimentary rocks. Feldspar crystals are commonly twinned and pale coloured and are readily altered to clay minerals.

Members of the feldspathoid group have an internal structure and chemistry close to those of feldspars, being sodium, potassium and calcium aluminium silicates, but contain less silica. Typical feldspathoids include nepheline, sodalite and lazurite.

Zeolites often occur in vesicles (cavities) in basic igneous rocks. They are aluminium silicates with water in their chemical structure (hydrated). This water can be removed by heating and taken up again when conditions are right, a property called reversible dehydration. Zeolites contain metals, especially sodium, calcium and potassium. Analcime, heulandite, natrolite and stilbite are all examples of zeolites.

Olivine

Chemical composition: $Fe^{2+}_2SiO_4$ (fayalite) to Mg_2SiO_4 (forsterite) **Crystal system:** Orthorhombic **Mineral habit:** Crystals commonly thick and tabular, with wedge-shaped terminations; usually granular, massive, compact **Cleavage:** Imperfect **Fracture:** Conchoidal **Hardness:** 7 **Specific gravity:** 4.39 (fayalite), 3.27 (forsterite) **Colour:** Green, greenish-yellow, yellow, yellowish-brown, brown, white, colourless **Streak:** Colourless **Lustre:** Vitreous to greasy; transparent to translucent

▼ Olivine, variety peridot, from St. John's Island, Red Sea, Egypt.

Olivine is the general name for a series of minerals ranging in composition from iron silicate to magnesium silicate. Fayalite, is the iron-rich end member of the series and at the other end is magnesium-rich forsterite. Both end-members of the series have near identical properties, including hardness and crystal system, but their differing chemistry gives the minerals a different specific gravity. Members of this series form as primary minerals in igneous rocks and are an important constituent of the Earth's upper mantle. Fayalite also occurs in metamorphosed iron-bearing sediments, whilst forsterite is found in basic igneous rocks, especially basalts, and in ultrabasic rocks, notably peridotites and in metamorphosed dolomites. Dunite is an ultrabasic rock composed almost entirely of olivine. The green gem variety of olivine is called peridot. The colour of this gem varies in accordance with the chemical composition of the mineral.

▲ Olivine from Mount Franklin, Victoria, Australia.

Garnet group

Chemical composition: The name garnet refers to a complex group of aluminium silicates, three of which are included here: almandine, $Fe^{2+}_3Al_2(SiO_4)_3$, grossular, $Ca_3Al_2(SiO_4)_3$, pyrope, $Mg_3Al_2(SiO_4)_3$ **Crystal system:** Cubic **Mineral habit:** Almandine: crystals dodecahedral, trapezohedral; also massive, granular, compact; grossular: crystals dodecahedral, trapezohedral; also massive, granular, compact; pyrope: crystals rare, dodecahedral, trapezohedral; commonly occurs in massive or granular habit **Cleavage:** None **Fracture:** Uneven to conchoidal **Hardness:** Grossular: 6½ to 7; almandine; pyrope 7½ **Specific gravity:** Almandine: 4.32; grossular: 3.59; pyrope: 3.58 **Colour:** Almandine: deep red, brownish red, brownish black; grossular: colourless, white, yellow, green, orange, red, brown; pyrope: pink, purplish, red, orange-red, black **Streak:** White **Lustre:** Vitreous to resinous; transparent to translucent

▼ Almandine from Langbanshyttan, Sweden.

The garnet group of aluminium silicate minerals occurs in metamorphic and igneous rock. Almandine and grossular form in schists and marbles, while pyrope tends to be found in ultrabasic igneous rocks such as peridotites and in metamorphic serpentinites. Because of their hardness and resistance to erosion, some worn garnet crystals occur in placer sands and gravels. Various forms of garnet have been used as gemstones, their intense colour and hardness making them valuable for this purpose. There are three other well-known garnets: spessartine, andradite and uvarovite.

▲ Spessartine garnet, from Broken Hill, New South Wales, Australia.

▶ Grossular from
Asbestos, Richmond
County, Quebec,
Canada.

◀ Garnet.

▶ Pyrope from
Elie Ness, Fife,
Scotland, UK.

Chondrodite

Chemical composition: $Mg_5(SiO_4)_2F_2$ **Crystal system:** Monoclinic **Mineral habit:** Crystals variable and modified, commonly twinned; also massive **Cleavage:** Indistinct **Fracture:** Uneven to subconchoidal **Hardness:** 6 to 6½ **Specific gravity:** 3.16 to 3.26 **Colour:** Yellow, orange, red, brown **Streak:** Greyish-white **Lustre:** Vitreous; transparent to translucent

▼ Chondrodite with spinel in marble from Sungate Mine, Lục Yên District, Vietnam.

Chondrodite usually occurs in contact metamorphosed limestones and dolomites, and rarely in calcium-rich igneous rocks called carbonatites. This mineral can be dissolved in hot hydrochloric acid, and when this cools down, a gelatinous precipitate forms. It does not melt when held in a flame. Chondrodite belongs to the humite group of silicate minerals, which includes clinohumite, norbergite and humite.

Topaz

Chemical composition: $Al_2SiO_4F_2$ **Crystal system:** Orthorhombic **Mineral habit:** Crystals prismatic, frequently highly modified, sometimes very large; also massive, granular, columnar **Cleavage:** Perfect **Fracture:** Subconchoidal to uneven **Hardness:** 8 **Specific gravity:** 3.40 to 3.60 **Colour:** Colourless, white, blue, greenish, yellowish, yellowish-brown, orange, grey, pale purple, pink **Streak:** Colourless **Lustre:** Vitreous; transparent to translucent

▼ Topaz from India.

Topaz is mainly found in rhyolitic and granitic pegmatites, where it can occur as very large crystals, in cavities and in quartz veins formed at high temperatures. The crystals can be up to 300 kg (660 lb) in weight. Topaz also occurs in rocks altered by contact metamorphism and in placer deposits. Associated minerals include quartz, tourmaline, beryl, fluorite, cassiterite, albite, microcline and muscovite. Topaz is used as a gemstone. It defines point 8 on the Mohs hardness scale. It does not dissolve in acids and it cannot be melted in a flame.

Willemite

Chemical composition: Zn_2SiO_4 **Crystal system:** Trigonal **Mineral habit:** Crystals prismatic, hexagonal; also massive, fibrous, compact, granular **Cleavage:** Poor **Fracture:** Conchoidal to uneven **Hardness:** 5½ **Specific gravity:** 3.89 to 4.19 **Colour:** Colourless, white, green, grey, red, brown, yellow **Streak:** Colourless **Lustre:** Vitreous to resinous; transparent to translucent

Often found as a secondary mineral in the altered parts of zinc-ore deposits, especially those containing sphalerite, willemite can also form in metamorphosed limestones. Rarely it occurs as a primary mineral and is mined for its zinc content. This mineral fluoresces an intense yellow or green when placed under different wavelengths of ultraviolet light and it can be phosphorescent. If powdered, it will dissolve in hydrochloric acid, but it cannot be melted.

▼ Willemite from Franklin, New Jersey, USA.

Phenakite

Chemical composition: Be_2SiO_4 **Crystal system:** Trigonal **Mineral habit:** Crystals prismatic, rhombohedral, acicular, commonly twinned; also granular, columnar, as radiating rounded masses **Cleavage:** Distinct **Fracture:** Conchoidal **Hardness:** 7½ to 8 **Specific gravity:** 2.96 to 3.00 **Colour:** Colourless, pink, pinkish-red, brown, yellow **Streak:** White **Lustre:** Vitreous; transparent

Phenakite forms in granitic pegmatites, greisens, mica schists and some mineral veins. It occurs with quartz, beryl, topaz, chrysoberyl and apatite. Phenakite will not melt in a flame or dissolve in acids. This mineral has been used as a gemstone, and its hardness of 7½ to 8 provides good resistance to wear. When facetted, it has a brilliant lustre not unlike that of diamond.

◄ Phenakite from Momeik, Myanmar.

Staurolite

Chemical composition: $Fe^{2+}_2Al_9Si_4O_{23}(OH)$
Crystal system: Monoclinic   **Mineral habit:**
Crystals short prismatic, often twinned,
commonly cruciform (cross-shaped), at 90° or
60°; also granular   **Cleavage:** Distinct   **Fracture:**
Uneven to subconchoidal   **Hardness:** 7 to 7½
Specific gravity: 3.74 to 3.83   **Colour:** Dark
brown, red-brown, yellowish-brown, brownish-
black   **Streak:** Colourless to grey   **Lustre:**
Vitreous to resinous; translucent to opaque

▶ Staurolite
from Litchfield,
Litchfield County,
Connecticut, USA.

Staurolite is found mainly in regionally
metamorphosed rocks such as gneisses and mica
schists, in which it can occur as porphyroblasts.
Associated minerals include kyanite, garnet,
muscovite and quartz. Staurolite's hardness makes
it resistant to erosion and so it also accumulates in
placer sands. The cruciform structure of staurolite
is well known, its name being taken from the
Greek 'stauros', meaning cross.

Zircon

Chemical composition: $Zr(SiO_4)$ **Crystal
system:** Tetragonal   **Mineral habit:** Crystals
prismatic, terminations dipyramidal; occasionally
as sheaf-like and radiating fibrous aggregates,
rarely granular   **Cleavage:** Imperfect   **Fracture:**
Uneven   **Hardness:** 7½   **Specific gravity:** 4.60
to 4.70   **Colour:** Colourless, brown, reddish,
purple, yellow, green, greyish   **Streak:** White
Lustre: Vitreous to adamantine; transparent

▼ Zircon from
Seiland Island, Alta,
Finnmark, Norway.

Zircon usually forms as an accessory mineral in
igneous rocks, including granites, pegmatites
and syenites, and in some metamorphic rocks,
such as gneisses and schists. It occurs in detrital
sedimentary rocks and beach sands, through
the weathering, erosion and redeposition of
its primary source. This mineral is also found in
rocks on the moon and in meteorites. Zircon is
insoluble in acids and does not melt in a flame.
It is often weakly radioactive, because it can
contain small traces of uranium. This property
is used in radiometric dating of rocks. Zircon is
the main ore of zirconium. Some varieties are
used as gemstones, the specimens usually being
recovered from detrital pebbles in placer deposits.

Andalusite

Chemical composition: Al_2SiO_5 **Crystal system:** Orthorhombic **Mineral habit:** Crystals prismatic, with square or cruciform cross-section; also massive, compact and as columnar and fibrous aggregates **Cleavage:** Distinct **Fracture:** Uneven to subconchoidal **Hardness:** 6½ to 7½ **Specific gravity:** 3.13 to 3.21 **Colour:** Reddish-brown, pink, red, white, grey, greenish, yellowish **Streak:** Colourless **Lustre:** Vitreous; transparent to opaque

Andalusite generally forms in metamorphic rocks, including slates, schists and gneisses. Associated minerals include kyanite, corundum, cordierite and sillimanite. Frequently occurring in the low-grade metamorphic rock, slate, a variety of andalusite called chiastolite has cruciform (cross-shaped) crystals. Andalusite is occasionally found in granites and pegmatites. It has the same chemical composition as sillimanite and kyanite, but a different internal structure. Insoluble and unable to be melted in a flame, this mineral sometimes exhibits pleochroism, appearing pink when viewed from one direction and green when viewed from another.

▲ Andalusite var. chiastolite from Bimbowrie, South Australia.

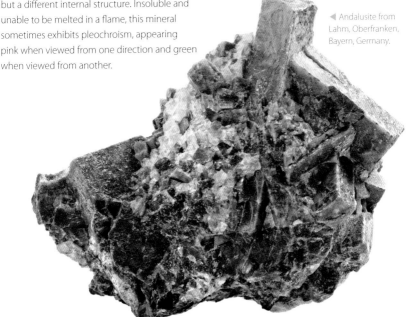

◄ Andalusite from Lahm, Oberfranken, Bayern, Germany.

Sillimanite

Chemical composition: Al_2SiO_5 **Crystal system:** Orthorhombic **Mineral habit:** Crystals prismatic, square-sectioned, vertically striated; usually massive, fibrous, columnar **Cleavage:** Perfect **Fracture:** Uneven **Hardness:** 6½ to 7½ **Specific gravity:** 3.23 to 3.27 **Colour:** Colourless, white, grey, yellowish, greenish, bluish, brown **Streak:** Colourless **Lustre:** Vitreous to silky; transparent to translucent

▶ Sillimanite.

Sillimanite is trimorphous with kyanite and andalusite. These minerals share the same chemical formula but have a different internal structure. Sillimanite occurs in metamorphic rocks such as gneisses and schists and in the igneous rock, granite. Associated minerals include corundum, cordierite and andalusite. It has also been found as inclusions in feldspars and quartz. Unusual pale blue-grey translucent crystals are sometimes used as gemstones and may show chatoyancy.

Kyanite

Chemical composition: Al_2SiO_5 **Crystal system:** Triclinic **Mineral habit:** Crystals flattened, bladed, often twisted or bent; also massive, fibrous **Cleavage:** Perfect **Fracture:** Uneven **Hardness:** 5½ to 7 **Specific gravity:** 3.53 to 3.67 **Colour:** Commonly blue, grey, white or colourless; rarely yellowish, green, pink, orange, black **Streak:** Colourless **Lustre:** Vitreous to pearly; transparent to translucent

◀ Kyanite from Porteirinha, Minas Gerais, Brazil.

Is trimorphous with andalusite and sillimanite, kyanite has the same chemical composition. It commonly occurs in metamorphic rocks, especially schists and gneisses. Kyanite is also found in granites and pegmatites. Its varying hardness depends on the direction in which a scratching test is carried out. The lower figure is obtained when testing along the length of a crystal; the higher figure across a crystal face. Though typically blue, kyanite can vary considerably in colour. This mineral does not dissolve in acids or melt in a flame.

Titanite (Sphene)

Chemical composition: $CaTiSiO_5$ **Crystal system:** Monoclinic **Mineral habit:** Crystals flattened, wedge-shaped or prismatic, commonly twinned; also massive, compact **Cleavage:** Distinct **Fracture:** Subconchoidal **Hardness:** 5 to 5½ **Specific gravity:** 3.48 to 4.60 **Colour:** Colourless, green, brown, grey, yellow, red, black **Streak:** White **Lustre:** Adamantine to resinous; transparent to opaque

Named for its titanium content, titanite commonly forms as an accessory mineral in igneous rocks, especially syenites. It can also be found in various metamorphic rocks, including gneisses and schists.

When occurring in sufficient amounts, titanite is used as an ore of titanium. This mineral can be melted in a flame, producing a yellow glass, and it dissolves in sulphuric acid. An alternative name for titanite is sphene.

▶ Titanite from Maggia Valley, Ticino, Switzerland.

Dumortierite

Chemical composition: $(Al,Fe^{3+})_7(SiO_4)_3(BO_3)O_3$ **Crystal system:** Orthorhombic **Mineral habit:** Crystals rare, prismatic; usually massive, fibrous, granular, columnar **Cleavage:** Good **Fracture:** Uneven **Hardness:** 7 to 8 **Specific gravity:** 3.21 to 3.41 **Colour:** Blue, violet, pink, brown **Streak:** White **Lustre:** Vitreous to dull; transparent to translucent

Dumortierite occurs in metamorphic rocks rich in aluminium, and sometimes in pegmatites. It cannot be melted and will not dissolve when placed in acids. This widespread mineral is very hard, but its fibrous structure makes it unsuitable

▼ Dumortierite in quartz from Sahatany, Antsirabe, Madagascar.

for use as a gemstone, although quartz containing dumortierite crystals, is sometimes polished and used in jewellery.

Eudialyte

Chemical composition: $Na_{15}Ca_6Fe_3Zr_3Si(Si_{25}O_{73})$ $(O,OH,H_2O)_3(Cl,OH)_2$ **Crystal system:** Trigonal **Mineral habit:** Crystals tabular, prismatic, rhombohedral; also massive, granular **Cleavage:** Indistinct **Fracture:** Uneven **Hardness:** 5 to 6 **Specific gravity:** 2.74 to 3.10 **Colour:** Brownish-yellow, brown, red, pink **Streak:** Colourless **Lustre:** Vitreous to greasy or dull; translucent

▼ Eudialyte from Rasvumchorr Mt, Khibiny Massif, Murmansk Oblast, Russia.

Occurring chiefly in syenites and related igneous rocks, often in association with aegirine, nepheline and microcline feldspar, eudialyte is also found in pegmatites. As well as zirconium, some eudialyte contains yttrium and is a source of rare earth elements. It is a relatively rare mineral globally, but where found it can be locally abundant. Eudialyte readily dissolves in acids.

Euclase

Chemical composition: $BeAlSiO_4(OH)$ **Crystal system:** Monoclinic **Mineral habit:** Crystals prismatic, usually long but may be short and stout, frequently with striated faces; also massive, rarely as radial aggregates, fibrous **Cleavage:** Perfect **Fracture:** Conchoidal **Hardness:** 7½ **Specific gravity:** 2.99 to 3.10 **Colour:** White, blue, pale green, pale pinkish, pale yellow, colourless **Streak:** White **Lustre:** Vitreous; transparent to translucent

Euclase forms in low- to medium-grade, regionally metamorphosed rocks such as phyllites and schists, and in coarse-grained igneous pegmatites. It is also found in quartz-rich veins and placer sands. Euclase is occasionally cut as a gemstone. It does not dissolve in acids and can only be melted with difficulty when held in a flame.

▲ Euclase showing striated faces.

Datolite

Chemical composition: $CaB(SiO_4)(OH)$ **Crystal system:** Monoclinic **Mineral habit:** Crystals commonly short prismatic; also compact, granular **Cleavage:** None **Fracture:** Uneven to conchoidal **Hardness:** 5 to 5½ **Specific gravity:** 2.96 to 3.00 **Colour:** White, colourless, pale greenish, pale yellowish; impurities can cause the mineral to appear reddish, pink or brownish **Streak:** Colourless **Lustre:** Vitreous; transparent to translucent

Forming in veins and cavities in basic igneous rocks, where it occurs with zeolite minerals, calcite, quartz and prehnite, datolite is also found in metal-bearing mineral veins, and in cavities in gneisses, serpentinites and granites. It is soluble in acids and melts when heated, turning the flame green.

▼ Datolite from Berger's Quarry, West Paterson, New Jersey, USA.

Gadolinite-(Y)

Chemical composition: $Y_2Fe^{2+}Be_2O_2(SiO_4)_2$ **Crystal system:** Monoclinic **Mineral habit:** Crystals usually prismatic, rough, often flattened; commonly compact, massive **Cleavage:** None **Fracture:** Conchoidal **Hardness:** 6½ to 7 **Specific gravity:** 4.36 to 4.77 **Colour:** Black, brown, greenish-black, occasionally light green **Streak:** Greenish-grey **Lustre:** Vitreous to greasy; transparent to translucent

▼ Gadolinite in rock groundmass.

Gadolinite-(Y), together with its closely related series members gadolinite-(Ce) and gadolinite-(Nd), forms in igneous rocks, particularly those of acidic composition. It occurs especially in pegmatites and granites, with fluorite and allanite. It is also found in schists and other regionally metamorphosed rocks. This mineral contains the element yttrium, which is very similar to the lanthanide group of elements, part of the so-called 'rare earth elements' (REEs), which are extracted from gadolinite series minerals. Gadolinite is often slightly radioactive, due to impurities of uranium and can be dissolved in acids, but will not melt when held in a flame.

Zoisite

Chemical composition: $Ca_2Al_3[Si_2O_7][SiO_4]O(OH)$
Crystal system: Orthorhombic **Mineral habit:**
Crystals prismatic, commonly striated; also massive,
columnar, compact **Cleavage:** Perfect **Fracture:**
Uneven to conchoidal **Hardness:** 6 to 7
Specific gravity: 3.15 to 3.36 **Colour:** White,
grey, green, greenish-brown, greenish-grey;
pink (thulite); blue-purple (tanzanite) also;
colourless **Streak:** Colourless **Lustre:** Vitreous
to pearly; transparent to translucent

Zoisite is found in a great variety of rocks,
including metamorphic rocks such as gneisses,
schists and marbles and igneous pegmatites.
It is also found in quartz veins, where it can be
associated with sulphide minerals. The blue-
purple variety, tanzanite, the colour of which can
be enhanced by heating, is used as a gemstone.
The pink colouring of the variety thulite is caused
by traces of manganese. Zoisite is insoluble in
acids, but melts when held in a flame, becoming a
white-coloured glass.

▶ Zoisite from
Saualpe, Wolfsberg,
Carinthia, Austria.

▲ Zoisite var.
Tanzanite from
Merelani Hills,
Tanzania.

▶ Zoisite var. thulite
from Hindremseter,
Leksvik, Trøndelag,
Norway.

Epidote

Chemical composition: $Ca_2(Al_2Fe^{3+})[Si_2O_7][SiO_4]$ O(OH) **Crystal system:** Monoclinic **Mineral habit:** Crystals prismatic, commonly striated, tabular, acicular; also massive, granular, fibrous **Cleavage:** Perfect **Fracture:** Uneven **Hardness:** 6 **Specific gravity:** 3.38 to 3.49 **Colour:** Often yellowish-green to brownish-green, green, grey, black **Streak:** Colourless or greyish **Lustre:** Vitreous to pearly or resinous; transparent to opaque

Epidote occurs in metamorphosed igneous rocks as well as in amphibolites and gabbros. It is a common rock-forming mineral in schists and marbles. Epidote can also be produced by hydrothermal fluids affecting and altering feldspars, amphiboles and pyroxenes in igneous rocks. It displays pleochroism, appearing greenish when viewed from one direction and yellow when viewed from another. Epidote does not dissolve in acids, but can be melted in a flame.

▶ Epidote from Copper Mountain, Prince of Wales Island, Alaska, USA.

Allanite-(Ce)

Chemical composition: $CaCe(Al_2Fe^{2+})[Si_2O_7]$ $[SiO_4]O(OH)$ **Crystal system:** Monoclinic **Mineral habit:** Crystals tabular, prismatic, acicular, bladed, commonly twinned; also massive, compact, granular **Cleavage:** None **Fracture:** Conchoidal to uneven **Hardness:** 5½ to 6 **Specific gravity:** 3.50-4.20 **Colour:** Very dark brown to black **Streak:** Greyish brown **Lustre:** Resinous to submetallic; translucent to opaque

▼ Allanite from Luzenac, Ariege, France.

Allanite-(Ce), a member of the epidote group, forms as an accessory mineral in various igneous rocks, notably in granites and acidic pegmatites. It can also occur in high grade metamorphic rocks such as gneisses. Allanite may contain rare earth elements within its chemical structure. The cerium (Ce) in the name and formula may be replaced by lanthanum (La), neodymium (Nd) or yttrium (Y). Other elements that can occur in allanite include uranium and thorium, and if they are present the mineral will be radioactive. When heated in a flame, allanite melts and expands, becoming a black magnetic glass. It can be dissolved in hydrochloric acid.

Hemimorphite

Chemical composition: $Zn_4Si_2O_7(OH)_2.H_2O$ **Crystal system:** Orthorhombic **Mineral habit:** Crystals usually tabular, striated, doubly terminated crystals exhibiting hemimorphism; usually massive, mammillary, stalactitic, granular **Cleavage:** Perfect **Fracture:** Subconchoidal to uneven **Hardness:** 4½ to 5 **Specific gravity:** 3.48 **Colour:** Colourless, white, blue, green, yellowish, grey, brown **Streak:** Colourless **Lustre:** Vitreous to silky or dull; transparent to translucent

Hemimorphite forms by the alteration of primary minerals containing zinc. Associated minerals include sphalerite, galena, cerussite, smithsonite, calcite, aurichalcite and anglesite. Hemimorphite will dissolve in concentrated acids, producing a gelatinous precipitate. When heated, water is given off, but hemimorphite can only be melted with difficulty. The mineral's name is derived from its hemimorphism (the property of displaying different shapes at the opposing ends of doubly terminated crystals).

▶ Hemimorphite from Roughton Gill, Caldbeck Fells, Cumbria, England, UK.

Axinite group

Chemical composition: $Ca_2Fe^{2+}Al_2BSi_4O_{15}OH$
(ferro-axinite) **Crystal system:** Triclinic **Mineral
habit:** Crystals sharp-edged; rarely massive
Cleavage: Good **Fracture:** Uneven to conchoidal
Hardness: 6½ to 7 **Specific gravity:** 3.25 to 3.28
Colour: Yellow, brown, reddish-brown, grey,
bluish **Streak:** Colourless **Lustre:** Vitreous;
transparent to subtranslucent

Axinite is the name of a group of minerals. The best-
known, axinite-(Fe) (or ferro-axinite), is dealt with
here. It can be found in marbles, where limestone
has been altered by contact metamorphism. Axinite
is a famous product of alpine metamorphism and
can also occur in granites. In addition to axinite-(Fe),
the group contains axinite-(Mg), with magnesium
in place of iron; axinite-(Mn), which contains
manganese, and the closely related manganese
bearing mineral tinzenite. Axinite melts easily in a
flame and gelatinizes in heated acids.

▼ Axinite from
Le Bourg-d'Oisans,
Isère, France.

Vesuvianite

Chemical composition: $(Ca,Na)_{19}(Al,Mg,$
$Fe)_{13}(SiO_4)_{10}(Si_2O_7)_4(OH,F,O)_{10}$ **Crystal system:**
Tetragonal **Mineral habit:** Crystals prismatic,
pyramidal; usually massive, columnar, granular
Cleavage: Indistinct **Fracture:** Uneven to
conchoidal **Hardness:** 6½ **Specific gravity:**
3.32 to 3.43 **Colour:** Green, brown, yellow,
red, purple, white, blue **Streak:** White **Lustre:**
Vitreous to resinous; transparent to translucent

Vesuvianite (also called idocrase) occurs in
metamorphic and igneous rocks, especially in
marbles and in syenites, as well as in ultrabasic rocks.
It is often found with a variety of other minerals,
including epidote, garnet, diopside, calcite, micas
and wollastonite. Vesuvianite melts quite easily,
producing a green or brown glass, but it is almost
insoluble in acids. Cyprine, which is usually blue in
colour, has recently been upgraded from a variety
of vesuvianite to a distinct mineral in its own right.

▶ Vesuvianite from
Drammen, Buskerud,
Norway.

Beryl

Chemical composition: $Be_3Al_2Si_6O_{18}$ **Crystal system:** Hexagonal **Mineral habit:** Crystals prismatic, often with pyramidal terminations, commonly striated, sometimes of great size; also columnar, compact **Cleavage:** Indistinct **Fracture:** Uneven to conchoidal **Hardness:** 7½ to 8 **Specific gravity:** 2.63 to 2.92 **Colour:** Colourless, white, green, yellowish-green, yellow, pink, red, blue, greenish-blue **Streak:** White **Lustre:** Vitreous; transparent to translucent

Beryl occurs in granites and pegmatites and also in greisens, schists and hydrothermal veins. Single crystals up to 6 m (20 ft) in length and 25 tonnes in weight have been found. It is much used as a gemstone and various colour varieties are named. Heliodor is a yellow variety; aquamarine is greenish blue or pale blue; emerald is the green form, though it can vary in depth of colour; morganite is pink; goshenite is the name given to the clear, colourless form and red beryl is the very rare deep red-pink variety. Beryl is insoluble in acids.

▶ Beryl var. heliodor from Agua Limpa, Minas Gerais, Brazil.

▲ Beryl var. aquamarine from Spitzkopje, Erongo Region, Namibia.

◀ Beryl var. emerald from Muzo Mine, Boyacá Department, Colombia.

▼ Red beryl from Thomas Mountains, Utah, USA.

◀ Beryl var. morganite from San Diego County, California, USA.

▶ Beryl var. goshenite from Goshen, Hampshire, Massachusetts, USA.

Tourmaline group

Chemical composition: Na(Mg,Fe,Li,Mn, Al)$_3$Al$_6$(BO$_3$)$_3$Si$_6$O$_{18}$(OH,F)$_4$ **Crystal system:** Trigonal **Mineral habit:** Crystals prismatic, acicular, usually striated, sometimes with triangular cross-section; also massive, compact, granular, fibrous **Cleavage:** Indistinct **Fracture:** Uneven to conchoidal **Hardness:** 7 **Specific gravity:** 2.90 to 3.10 **Colour:** Green, yellow-green, brown, black, pink, red, often zoned in multicolours **Streak:** Colourless **Lustre:** Vitreous to resinous; transparent, translucent, opaque

▼ Schorl.

Tourmaline is a group of closely related and chemically very complex minerals which commonly form in granites, granitic pegmatites and metamorphic rocks. Crystals can sometimes reach sizes of over one metre (39 in). The tourmalines occur with many minerals including feldspars, beryl, zircon and quartz. Chemically, schorl is rich in iron and is black, while dravite contains much magnesium and is usually brown; the mineral elbaite is rich in lithium and can have various colours including pink, green, blue or yellow. When tourmaline is pink and of a quality good enough to be facetted as a gemstone, it is given the name rubellite. Varieties that show a multicolour change from pinkish red to green are often referred to as 'watermelon tourmaline'.

▲ Tourmaline var. rubellite from Shaitanka, Sverdlovsk Oblast, Russia.

▶ Elbaite from Grotta d'Oggi, San Piero in Campo, Elba, Italy.

Dioptase

Chemical composition: $CuSiO_3.H_2O$ **Crystal system:** Trigonal **Mineral habit:** Crystals prismatic, often with rhombohedral terminations; also massive and as aggregates **Cleavage:** Perfect **Fracture:** Uneven to conchoidal **Hardness:** 5 **Specific gravity:** 3.28 to 3.35 **Colour:** Green to blue-green and turquoise blue **Streak:** Pale green-blue **Lustre:** Vitreous to greasy; transparent to translucent

Dioptase mainly forms where veins containing copper-rich minerals have been altered by oxidation. It can also be found in cavities in neighbouring rocks, often in association with native copper, cerussite, wulfenite, quartz and malachite. Though it is a brittle mineral, it is much valued by collectors because of its intense green colouring and fine crystal habit. Dioptase is soluble in hydrochloric acid and ammonia but does not melt in a flame.

▼ Dioptase from Tsumeb, Oshikoto Region, Namibia.

Ilvaite

Chemical composition: $CaFe^{3+}Fe^{2+}_2(Si_2O_7)$ $O(OH)$ **Crystal system:** Orthorhombic **Mineral habit:** Crystals prismatic, striated, diamond-shaped in cross-section; also massive, compact, columnar **Cleavage:** Distinct **Fracture:** Uneven **Hardness:** 5½ to 6 **Specific gravity:** 3.99 to 4.05 **Colour:** Black, grey-black **Streak:** Black, brownish, greenish **Lustre:** Submetallic; opaque

▼ Ilvaite from Chifeng, Inner Mongolia, China.

Occurring where rocks have been altered by contact metamorphism, both by intruded magma and extrusive lavas, ilvaite is also found in syenites and in ore deposits containing zinc, copper and iron minerals. This mineral is soluble in hydrochloric acid, producing gelatinization, and it melts readily in a flame.

Cordierite

Chemical composition: $Mg_2Al_4Si_5O_{18}$
Crystal system: Orthorhombic **Mineral habit:**
Crystals prismatic, with a rectangular cross-section;
usually massive, compact, granular **Cleavage:**
Distinct **Fracture:** Conchoidal **Hardness:** 7 to 7½
Specific gravity: 2.60 to 2.66 **Colour:** Blue,
violet-blue; grey, greenish, yellow, brown
Streak: Colourless **Lustre:** Vitreous; transparent
to translucent

▶ Cordierite from
Bjordam, Bamble,
Telemark, Norway.

Forming in thermally metamorphosed
aluminium-rich rocks, especially hornfels created
by metamorphism of clay-rich sedimentary
rocks, often cordierite is associated with quartz,
andalusite, biotite, spinel, garnet and sillimanite.
It can also be found in various igneous rocks,
including granites, pegmatites and andesites, and
in placer deposits. Cordierite is often pleochroic,
appearing blue when viewed from some angles
and yellowish grey from others. When placed in a
flame, only the thin edges of a specimen will melt,
and the mineral is insoluble in acids.

Benitoite

Chemical composition: $BaTiSi_3O_9$ **Crystal
system:** Hexagonal **Mineral habit:** Crystals
pyramidal, tabular, faces often triangular; also
granular **Cleavage:** Indistinct **Fracture:**
Conchoidal to uneven **Hardness:** 6 to 6½
Specific gravity: 3.65 **Colour:** Blue, purple,
pinkish, colourless, white **Streak:** Colourless
Lustre: Vitreous; transparent to translucent

Benitoite occurs in serpentinites and schists and
in placer sands. It is a rare mineral, found with
neptunite, albite and natrolite. When placed
under ultraviolet light, it fluoresces blue. It is a
pleochroic mineral, appearing blue or colourless
depending on the angle from which it is viewed.

◀ Benitoite from
San Benito County,
California, USA.

Enstatite

Chemical composition: $MgSiO_3$ **Crystal system:** Orthorhombic **Mineral habit:** Crystals prismatic; usually massive, fibrous or lamellar **Cleavage:** Good **Fracture:** Uneven **Hardness:** 5 to 6 **Specific gravity:** 3.20 to 3.90 **Colour:** Colourless, olive green, greenish- or yellowish-white, brown, grey **Streak:** Colourless, greyish **Lustre:** Vitreous to pearly; transparent to opaque

Enstatite belongs to the pyroxene group of minerals and is a common constituent of basic and ultrabasic igneous rocks, including basalts, gabbros and peridotites. It is also found in metamorphic rocks, especially those of high grade, and in both iron and stony meteorites. Enstatite is far less common as well-formed prismatic crystals in cavities. It is insoluble and can be melted only with extreme difficulty. Bronzite is an iron-rich variety of enstatite.

▲ Enstatite from Bamble, Telemark, Norway.

▶ Enstatite var. bronzite from Webster, Jackson, North Carolina, USA.

Diopside

Chemical composition: $CaMgSi_2O_6$ **Crystal system:** Monoclinic **Mineral habit:** Crystals prismatic; also massive, columnar, lamellar, granular **Cleavage:** Good **Fracture:** Uneven to conchoidal **Hardness:** 5½ to 6½ **Specific gravity:** 3.22 to 3.38 **Colour:** Colourless, white, grey, pale to dark green, brown **Streak:** White, greyish **Lustre:** Vitreous to dull; transparent to opaque

▼ Diopside in calcite from Jungaluk, Sar-e Sang, Badakhshan, Afghanistan.

Diopside belongs to the pyroxene group, and is the magnesium-rich end member of the diopside-hedenbergite series. It is found in metamorphic rocks, especially those rich in calcium, and in certain basic and ultrabasic igneous rocks, including basalts and gabbros. Diopside does not dissolve in acids, but can be melted with difficulty in a flame, producing a green glass.

Hedenbergite

Chemical composition: $CaFe^{2+}Si_2O_6$ **Crystal system:** Monoclinic **Mineral habit:** Crystals prismatic; also commonly massive, lamellar **Cleavage:** Good **Fracture:** Uneven to conchoidal **Hardness:** 5½ to 6½ **Specific gravity:** 3.56 **Colour:** Greenish, brownish, greyish-black, black **Streak:** White, greyish **Lustre:** Vitreous to resinous or dull; translucent to opaque

Hedenbergite is the iron-rich end member of the diopside-hedenbergite series within the pyroxene group of minerals. It forms where limestones have been altered to marbles by contact metamorphism and in metamorphic rocks rich in iron. Associated minerals include calcite, epidote and garnet. This mineral is also found in igneous rocks including granites and syenites and in meteorites. Hedenbergite is insoluble in acids, but melts readily in a flame, producing a black magnetic glass.

▲ Hedenbergite from Zillertal, Tyrol, Austria.

Augite

Chemical composition: $(Ca,Mg,Fe)_2Si_2O_6$
Crystal system: Monoclinic **Mineral habit:**
Crystals prismatic, commonly twinned; also
massive, granular, compact **Cleavage:** Good
Fracture: Uneven to conchoidal **Hardness:**
5½ to 6 **Specific gravity:** 3.19 to 3.56 **Colour:**
Greenish, black, brownish, purplish-brown
Streak: Greyish-green **Lustre:** Vitreous to dull;
translucent to opaque

Augite occurs as a widespread rock-forming
member of the pyroxene group of minerals. It is
common in basic igneous rocks, especially basalts
and gabbros, in which it may make up nearly
half the rock's composition. Augite also occurs in
ultrabasic and intermediate rocks and in some
metamorphic rocks. As with other pyroxene
minerals, there are two cleavage planes through
the crystals, which are nearly at right angles to each
other. This is a good distinguishing feature when

▼ Augite from Fassa
Valley, Trentino, Italy.

comparing augite to members of the similarly
coloured hornblende group, which have 120°
between their cleavage planes. Augite is insoluble
in acids and can only be melted with the greatest
difficulty.

Aegirine

Chemical composition: $NaFe^{3+}Si_2O_6$ **Crystal
system:** Monoclinic **Mineral habit:** Crystals
prismatic, striated, acicular, commonly twinned;
also as tufts, or felted aggregates **Cleavage:**
Good **Fracture:** Uneven **Hardness:** 6
Specific gravity: 3.50 to 3.60 **Colour:** Dark
green, brown, black **Streak:** Yellowish-grey
Lustre: Vitreous to resinous; translucent
to opaque

Aegirine, also known as acmite, forms in
intermediate igneous rocks including syenites. It is
also found in carbonatites and some metamorphic
rocks, including schists and gneisses. This mineral
is a pyroxene, related to augite. Aegirine is
insoluble in acids, but melts easily, colouring a
flame yellow because of its sodium content.

▼ Aegirine from
Magnet Cove, Hot
Spring County,
Arkansas, USA.

Spodumene

Chemical composition: $LiAlSi_2O_6$ **Crystal system:** Monoclinic **Mineral habit:** Crystals prismatic, often flattened, striated, twinned, sometimes very large; also massive **Cleavage:** Perfect **Fracture:** Uneven to subconchoidal **Hardness:** 6½ to 7 **Specific gravity:** 3.10 to 3.20 **Colour:** Colourless, white, greyish, yellow, green, pink, lilac **Streak:** White **Lustre:** Vitreous to dull; transparent to translucent

▶ Spodumene var. hiddenite from Alexander County, North Carolina, USA.

Two colour forms of spodumene, a pyroxene mineral, have been named as gemstone varieties: the green examples are called hiddenite, and the well-known lilac-coloured examples are kunzite. Spodumene is pleochroic, appearing to have different colours, depending on the angle from which the crystal is viewed. It forms in pegmatites, occurring with quartz, feldspars, micas, tourmaline, topaz and beryl. Spodumene is often altered to clay and mica. Crystals over 15 m (49 ft) long and weighing over 60 tonnes have been found. Spodumene is insoluble in acids, but melts in a flame, colouring the flame red because of its lithium content. It has been mined for this metal, which is important for making batteries.

◀ Spodumene from Chesterfield, Massachusetts, USA.

▼ Kunzite from Afghanistan.

Jadeite

Chemical composition: $NaAlSi_2O_6$ **Crystal system:** Monoclinic **Mineral habit:** Crystals prismatic, commonly twinned and striated, rare; usually granular, massive **Cleavage:** Good **Fracture:** Uneven **Hardness:** 6 **Specific gravity:** 3.25 to 3.35 **Colour:** Greenish, colourless, white, grey, purple **Streak:** Colourless **Lustre:** Vitreous to greasy; transparent to translucent

Jadeite is a member of the pyroxene group, occurring in ultrabasic igneous rocks that have been altered by serpentinization, and in schists. It is used ornamentally as a form of jade. When heated, it melts quite readily, producing a transparent globule, but it will not dissolve in acids.

▶ Jadeite.

Hornblende series

Chemical composition: $Ca_2(Fe,Mg)_4Al(Si_7Al)$ $O_{22}(OH)_2$ **Crystal system:** Monoclinic **Mineral habit:** Crystals prismatic, commonly twinned, often with nearly hexagonal cross-sections; also massive, columnar, granular, bladed, fibrous **Cleavage:** Perfect **Fracture:** Uneven to subconchoidal **Hardness:** 5 to 6 **Specific gravity:** 3.00 to 3.40 **Colour:** Green, greenish-brown, black **Streak:** White **Lustre:** Vitreous; translucent to opaque

▼ Hornblende from Madagascar.

Hornblende is an informal name for a series of minerals, varying from ferro-hornblende to magnesio-hornblende. These belong to the amphibole group of minerals, which are common rock-formers. They are superficially similar to the pyroxene group, but hornblende has cleavage planes intersecting at 120°, whereas pyroxenes cleave at nearly 90°. This mineral occurs mainly in acid igneous rocks, but is also found in intermediate and ultrabasic rocks. Hornblende can form in certain high-grade metamorphic rocks. It will not dissolve in acids, but melts in a flame, producing a green glass.

Glaucophane

Chemical composition: $Na_2Mg_3Al_2Si_8O_{22}(OH)_2$
Crystal system: Monoclinic **Mineral habit:**
Crystals prismatic, acicular, rare; usually massive,
columnar, fibrous, granular **Cleavage:** Perfect
Fracture: Uneven to conchoidal **Hardness:**
5 to 6 **Specific gravity:** 3.00 to 3.15 **Colour:**
Grey, blue-black, blue **Streak:** Greyish-blue
Lustre: Vitreous to dull or pearly; translucent

▼ Glaucophane from
Groix island, Morbihan,
Brittany, France.

Glaucophane belongs to the amphibole group of
minerals, and commonly forms in medium-grade
schists. It is often associated with muscovite,
chlorite, epidote, almandine and jadeite.
Glaucophane will not dissolve in acids, but melts
in a flame, producing a greenish glass.

Riebeckite

Chemical composition: $Na_2(Fe^{2+}_3Fe^{3+}_2)Si_8O_{22}(OH)_2$
Crystal system: Monoclinic **Mineral habit:**
Crystals prismatic, striated; also massive,
columnar, granular, fibrous, asbestiform
Cleavage: Perfect **Fracture:** Uneven
Hardness: 5 to 5½ **Specific gravity:** 3.28
to 3.44 **Colour:** Dark blue, black, grey,
brown **Streak:** Grey **Lustre:** Vitreous or silky;
translucent to opaque

Riebeckite occurs in a variety of igneous rocks,
especially granites, syenites and pegmatites. This
amphibole mineral is also found in banded ironstones,
metamorphic schists and gneisses. Associated
minerals include aegirine, albite, tremolite, magnetite,
hematite, calcite and quartz. Extremely fibrous
riebeckite (a chatoyant variety named crocidolite)
is sometimes called blue asbestos and is one of the
most dangerous forms of asbestos. Riebeckite is
insoluble in acids, but melts quite readily, colouring a
flame yellow because of its sodium content.

▶ Riebeckite crystals
in groundmass.

Actinolite

Chemical composition: $Ca_2(Mg,Fe^{2+})_5Si_8O_{22}(OH)_2$
Crystal system: Monoclinic **Mineral habit:**
Crystals bladed or short and stout, commonly
twinned; usually fibrous, radiating, columnar,
massive, granular **Cleavage:** Good **Fracture:**
Uneven to subconchoidal **Hardness:** 5 to 6
Specific gravity: 3.03 to 3.24 **Colour:** Light to
dark green **Streak:** White **Lustre:** Vitreous to
dull; transparent to opaque

Actinolite forms a series with the mineral tremolite
within the amphibole group. It occurs in a variety
of metamorphosed rocks, especially schists and
altered basic igneous rocks such as serpentinites.
It can also result from metamorphic changes to
pyroxene minerals. Fibrous actinolite is a form of
asbestos. Nephrite, a type of jade, is a compact
variety of actinolite or tremolite. Nephrite jade
is not as hard as jadeite jade, but has a similar
colour, ranging from white through to dark green.
Actinolite melts only with difficulty and does not
dissolve in acids.

▼ Nephrite from
Fraser River, British
Columbia, Canada.

▼ Actinolite
from Sutherland,
Scotland, UK.

Tremolite

Chemical composition: $Ca_2Mg_5Si_8O_{22}(OH)_2$
Crystal system: Monoclinic **Mineral habit:**
Crystals bladed or short and stout, often twinned;
usually fibrous, columnar, radiating, massive,
granular **Cleavage:** Good **Fracture:** Uneven
to subconchoidal **Hardness:** 5 to 6 **Specific
gravity:** 2.99 to 3.03 **Colour:** Colourless, white,
grey, green, pink, brownish **Streak:** White
Lustre: Vitreous; transparent to translucent

Tremolite is a member of the amphibole group of
minerals, closely related to actinolite, and is also
a component of nephrite jade. When it occurs
as fibres, it can be a form of asbestos. It is found
in metamorphosed limestones, especially those

▼ Tremolite
from Canaan,
Connecticut, USA.

containing magnesium, and in serpentinites.
Tremolite is insoluble in acids, and melts with
difficulty, producing a white-coloured glass.

Arfvedsonite

Chemical composition: $NaNa_2(Fe^{2+}_4Fe^{3+})$ $Si_8O_{22}(OH)_2$ **Crystal system:** Monoclinic **Mineral habit:** Crystals prismatic, frequently tabular; also in aggregates **Cleavage:** Perfect **Fracture:** Uneven **Hardness:** 5 to 6 **Specific gravity:** 3.30 to 3.50 **Colour:** Bluish-black, black **Streak:** Blue-grey **Lustre:** Vitreous; opaque

Arfvedsonite is a member of the amphibole group of minerals. It typically occurs in nepheline syenites, pegmatites, and in schists. This mineral will not dissolve in acids, but melts readily, producing a magnetic black glass.

◀ Arfvedsonite from Hurricane Mountain, New Hampshire, USA.

Rhodonite

Chemical composition: $Mn^{2+}SiO_3$ **Crystal system:** Triclinic **Mineral habit:** Crystals tabular, rare; usually massive, granular, compact **Cleavage:** Perfect **Fracture:** Conchoidal to uneven **Hardness:** 5½ to 6½ **Specific gravity:** 3.57 to 3.76 **Colour:** Pink, red, brownish red, often with black veins **Streak:** White **Lustre:** Vitreous to pearly; transparent to translucent

Rhodonite commonly occurs in manganese-rich rocks that have been altered by metasomatism, hydrothermal processes or metamorphism. It is occasionally found as large crystals, associated with willemite, calcite and franklinite. Rhodonite can be veined with dark markings due to the presence of manganese oxides. It is insoluble in acids, but melts in a flame, producing a brown or red glass. Rhodonite is often used ornamentally.

◀ Rhodonite from Siberia.

▶ Rhodonite from North Mine, Broken Hill, New South Wales, Australia.

Wollastonite

Chemical composition: CaSiO$_3$ **Crystal system:** Triclinic **Mineral habit:** Crystals tabular, often twinned; usually massive, granular, fibrous, compact **Cleavage:** Perfect **Fracture:** Splintery **Hardness:** 4½ to 5 **Specific gravity:** 2.86 to 3.09 **Colour:** White, grey; colourless, pale green **Streak:** White **Lustre:** Vitreous to pearly or silky; transparent to translucent

▼ Wollastonite from Jeffrey mine, Asbestos, Richmond County, Québec, Canada.

Wollastonite is a common mineral in marbles formed by contact metamorphism. It can produce attractive veining in these rocks and occurs with brucite, garnet, tremolite, diopside and epidote. Wollastonite also forms in some igneous rocks. It is soluble in strong acids and melts in a flame, producing a clear glass.

Pectolite

Chemical composition: NaCa$_2$Si$_3$O$_8$(OH) **Crystal system:** Triclinic **Mineral habit:** Crystals acicular, in spherical radiating aggregates, sometimes tabular; also massive **Cleavage:** Perfect **Fracture:** Splintery **Hardness:** 4½ to 5 **Specific gravity:** 2.84 to 2.90 **Colour:** White, colourless **Streak:** White **Lustre:** Vitreous to silky; transparent to translucent

Pectolite forms in vesicles in basic igneous rocks, especially fine-grained basalts, with zeolites, including analcime, chabazite, philipsite, heulandite and natrolite. It also occurs in some serpentinites and calcareous metamorphic rocks. When it is dissolved in hydrochloric acid, a siliceous gel is produced. If heated in a closed tube, water is given off, and the mineral colours a flame yellow, because of its sodium content.

▶ Pectolite from Copt Hill Quarry, Cowshill, Weardale, County Durham, England, UK.

Neptunite

Chemical composition: $KNa_2LiFe^{2+}_2Ti_2Si_8O_{24}$
Crystal system: Monoclinic **Mineral habit:**
Crystals prismatic, commonly with a square
cross-section **Cleavage:** Perfect **Fracture:**
Conchoidal **Hardness:** 5 to 6 **Specific gravity:**
3.19 to 3.23 **Colour:** Black, dark red-brown
Streak: Red-brown **Lustre:** Vitreous; opaque

Neptunite occurs as an accessory mineral in
intermediate igneous rocks such as syenites and
sometimes in serpentinites. It can be associated
with minerals such as eudialyte, natrolite and
benitoite. Neptunite cannot be dissolved in
hydrochloric acid, but will melt easily in a flame.

▲ Neptunite from
Gem Mine, San
Benito County,
California, USA.

Antigorite

Chemical composition: $Mg_3(Si_2O_5)(OH)_4$
Crystal system: Monoclinic **Mineral habit:**
Crystals very small, flaky; usually massive, granular,
lamellar, columnar, foliated, compact, fibrous
Cleavage: Perfect **Fracture:** Conchoidal, splintery
Hardness: 3½ to 4 **Specific gravity:** 2.50 to 2.60
Colour: White, greenish, yellowish, brownish,
bluish **Streak:** White **Lustre:** Resinous, greasy,
waxy, pearly, earthy; translucent to opaque

Antigorite belongs to the serpentine
subgroup of minerals. These are all minerals
which have the same internal structures
and similar chemistry. Antigorite forms in
serpentinites, derived from the alteration of
ultrabasic igneous rocks. It often occurs with
chrysotile and other fibrous minerals.

◀ Antigorite from
Hoher Eichham,
Prägraten, Tyrol, Austria.

Chrysotile

Chemical composition: $Mg_3(Si_2O_5)(OH)_4$ **Crystal
system:** Monoclinic **Mineral habit:** Commonly
massive or fibrous **Cleavage:** None **Fracture:**
Fibrous **Hardness:** 2½ **Specific gravity:** 2.53
Colour: Green, white, grey, yellow, brownish
Streak: White **Lustre:** Silky or greasy; translucent

Chrysotile, a member of the serpentine subgroup,
occurs in serpentinized rocks. It is a low-
temperature serpentine mineral, forming at
temperatures up to 250°C (482 °F), while antigorite
forms at higher temperatures. Chrysotile does not
melt in a flame, but dissolves in strong acids. Of all the
asbestos mined around the world, most is chrysotile.
When fibres or dust particles from asbestos are
breathed into the lungs, they can cause cancer.

▲ Chrysotile
in groundmass.

Talc

Chemical composition: $Mg_3Si_4O_{10}(OH)_2$
Crystal system: Triclinic **Mineral habit:**
Crystals tabular, thin; usually massive,
compact, foliated, fibrous, as globular
aggregates **Cleavage:** Perfect **Fracture:**
Uneven **Hardness:** 1 **Specific gravity:**
2.58 to 2.83 **Colour:** Green, grey, white,
brownish **Streak:** White **Lustre:** Pearly to
dull; translucent

▼ Talc from Koralpe,
Carinthia, Austria.

Talc represents point one on the Mohs scale
of hardness. It is very easily scratched with a
fingernail and has a greasy feel. Talc commonly
forms from the alteration of ultrabasic rocks and
the action of contact metamorphism on dolomitic
limestones. A compact form of talc is known by
the informal name of steatite or soapstone and
can be easily carved. Powdered talc is used as
'talcum powder'. This mineral cannot be melted in
a flame or dissolved in acids.

Chrysocolla

Chemical composition: $(Cu,Al)_2H_2Si_2O_5(OH)_4 \cdot nH_2O$ **Crystal system:** Orthorhombic **Mineral
habit:** Cryptocrystalline, botryoidal,
earthy **Cleavage:** None **Fracture:** Uneven
to conchoidal **Hardness:** 2½ to 3½ **Specific
gravity:** 1.93 to 2.40 **Colour:** Blue, blue-green,
green **Streak:** White **Lustre:** Vitreous, greasy,
earthy; translucent to opaque

Chrysocolla is a reasonably common mineral in
places where copper veins have been altered by
oxidation. It is often associated with malachite,
azurite, cuprite and other copper minerals. When
placed in a flame, a green colouring is produced,
which is indicative of copper. Chrysocolla
dissolves in hydrochloric acid, forming a siliceous
gel. The crystalline nature of chrysocolla is still
debated by scientists.

▶ Chrysocolla
from Santa Fe Mine,
Chiapas, Mexico.

Muscovite

Chemical composition: $KAl_2(AlSi_3O_{10})(OH)_2$
Crystal system: Monoclinic **Mineral habit:**
Crystals tabular, often with hexagonal cross-section, commonly twinned, sometimes
very large; also lamellar, scaly, compact,
massive **Cleavage:** Perfect **Fracture:**
Uneven **Hardness:** 2½ **Specific gravity:** 2.77
to 2.88 **Colour:** Colourless, grey, green,
yellowish, violet, brown, red **Streak:** Colourless
Lustre: Vitreous to pearly or silky; transparent
to translucent

Muscovite, often known as 'white mica', is a
common mineral in many rocks. It is abundant
in acid igneous rocks such as granites and
pegmatites and in regionally metamorphosed
schists and gneisses. Very large crystals, over
30 m (98 ft) across, have been found in some
pegmatites. Many sandstones contain small
flakes of detrital muscovite on their bedding
planes. Muscovite will not dissolve in acids and
only melts with difficulty. Fuchsite is a green-coloured chromium-rich variety and alurgite
is a rare manganese-rich variety, reddish
purple in colour.

▼ Muscovite and
Microlite from Brazil.

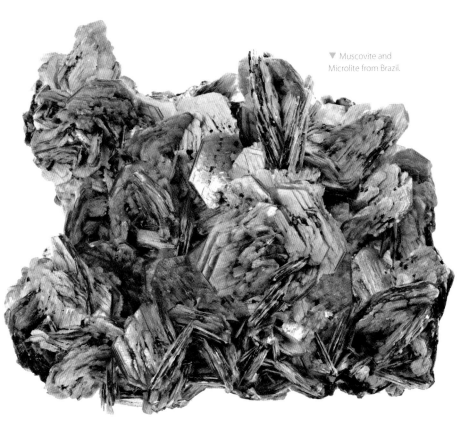

'Biotite'

Chemical composition: $KFe^{2+}_3(Al,Si_3O_{10})(OH)_2$
Crystal system: Monoclinic **Mineral habit:**
Crystals tabular, often with hexagonal
cross-section; usually scaly **Cleavage:**
Perfect **Fracture:** Uneven **Hardness:** 2½ to 3
Specific gravity: 3.30 **Colour:** Black, dark
brown, reddish-brown **Streak:** Brownish-
white **Lustre:** Submetallic, vitreous, pearly;
transparent to translucent

'Biotite' is not strictly a mineral, but an informal
name for dark-coloured micas that contain
significant iron. The formal name for the most
iron-rich micas with the properties shown above
is annite. However, the term 'biotite' is commonly

used in geology when the exact composition
of an iron-rich mica is not known. 'Biotite' occurs
in many igneous rocks, including granites,
pegmatites, diorites and gabbros, as a primary
constituent. It is also an important component of
many regionally metamorphosed rocks, especially
phyllites, schists and gneisses. In sedimentary
rocks it is a detrital mineral found on the bedding
surfaces of sandstones. Biotite can be melted
with difficulty, and dissolved only in heated
concentrated sulphuric acid.

Other micas include phlogopite, a brownish
mica that lacks iron in its chemical composition;
lepidolite, a pink or purple lithium-bearing variety
and glauconite, which contains sodium and gives a
green colouring to the sedimentary rock greensand.

▶ Annite from
Kalarsky Mountains,
Zabaykalsky Krai,
Russia.

▼ Biotite from
Switzerland.

▶ Phlogopite
from Sri Lanka.

Chamosite

Chemical composition: $(Fe^{2+},Mg,Al,Fe^{3+})_6(Si,Al)_4O_{10}(OH,O)_8$ **Crystal system:** Monoclinic **Mineral habit:** Massive, oolitic, compact, granular **Cleavage:** Good **Fracture:** Uneven **Hardness:** 3 **Specific gravity:** 3.00 to 3.40 **Colour:** Green, grey, black **Streak:** Greyish-green **Lustre:** Vitreous to dull; translucent to opaque

▼ Chamosite from Chamoson, Valais, Switzerland.

Chamosite is a member of the chlorite group of minerals. It is found in sedimentary rocks, especially ironstone deposits, and is associated with siderite, iron oxides and clay minerals such as kaolinite. When heated, it gives off water.

Clinochlore

Chemical composition: $Mg_5Al(Si_3Al)O_{10}(OH)_8$ **Crystal system:** Monoclinic **Mineral habit:** Crystals tabular, often with hexagonal cross-section; usually massive, foliated, granular, earthy **Cleavage:** Perfect **Fracture:** Uneven **Hardness:** 2 to 2½ **Specific gravity:** 2.60 to 3.02 **Colour:** Green, yellowish-green, blackish-green, white, pinkish-purple **Streak:** White **Lustre:** Vitreous to pearly; transparent to translucent

▲ Clinochlore from Achmatovsk Mine, Chelyabinsk Oblast, Russia.

Clinochlore belongs to the chlorite group. It generally forms in regionally metamorphosed rocks, including schists and serpentinites, and in igneous rocks, resulting from the alteration of pyroxenes, amphiboles and biotite by hydrothermal fluids. Clinochlore dissolves in concentrated acids. Instead of melting, it flakes apart when placed in a flame. The pinkish-purple variety of clinochlore is coloured by traces of chromium and is called kammererite.

Kaolinite

Chemical composition: $Al_2Si_2O_5(OH)_4$ **Crystal system:** Triclinic **Mineral habit:** Crystals small, hexagonal plates or scales; usually massive, compact, clayey, earthy **Cleavage:** Perfect **Fracture:** Uneven **Hardness:** 2 to 2½ **Specific gravity:** 2.63 **Colour:** Colourless, white, may be tinted yellowish, reddish, brown or bluish by impurities **Streak:** White **Lustre:** Pearly to dull or earthy; transparent to translucent

▼ Kaolinite from Soufrière Hills, Montserrat, Spain.

Kaolinite gives its name to a subgroup of related minerals. It is found, often in quantity, as a weathering and hydrothermal alteration product of feldspars and other aluminium silicates in granitic rocks. Kaolinite does not melt in a flame and will only dissolve in hot concentrated sulphuric acid. It can be of considerable economic importance as china clay.

Prehnite

Chemical composition: $Ca_2Al_2Si_3O_{10}(OH)_2$ **Crystal system:** Orthorhombic **Mineral habit:** Individual crystals rare, tabular, prismatic, pyramidal, often in barrel- or sheaf-shaped crystalline masses; usually granular, reniform, botryoidal, stalactitic **Cleavage:** Distinct **Fracture:** Uneven **Hardness:** 6 to 6½ **Specific gravity:** 2.80 to 2.95 **Colour:** Green, white, colourless, yellow, grey, pale blue **Streak:** White **Lustre:** Vitreous to pearly; transparent to translucent

Prehnite forms in a variety of igneous and metamorphic rocks, especially granites, diorites, gneisses and marbles as a secondary mineral. It also occurs in veins and cavities associated with hydrothermal activity, often with zeolite minerals, pectolite, datolite and calcite. When held in a flame, prehnite melts easily, creating a bubbly yellowish-white glass. It dissolves slowly in hydrochloric acid.

▶ Prehnite from Boylestone Quarry, Barrhead, Renfrewshire, Scotland, UK.

'Apophyllite'

Chemical composition: $KCa_4Si_8O_{20}(F,OH).8H_2O$
Crystal system: Tetragonal **Mineral habit:**
Crystals pseudocubic, prismatic, pyramidal,
tabular, often striated; also massive, granular,
lamellar **Cleavage:** Perfect **Fracture:** Uneven
Hardness: 4½ to 5 **Specific gravity:** 2.33 to
2.37 **Colour:** Colourless, white, pink, green
Streak: White **Lustre:** Vitreous to pearly;
transparent to translucent

▼ 'Apophyllite' from
Andreasberg, Harz,
Germany.

'Apophyllite' forms as a secondary mineral in
cavities in basaltic lavas with other minerals
such as analcime, calcite, prehnite and stilbite.
It has also been found in many rocks, including
granites, gneisses and limestones, as well as in
hydrothermal ore veins. When placed in a flame,
apophyllite melts easily, colouring the flame violet
owing to its potassium content. A gelatinous,
silica-rich globule is formed when it is dissolved
in hydrochloric acid. Strictly speaking, apophyllite
is a group name for several minerals including
fluorapophyllite-(K), hydroxyapophyllite-(K) and
fluorapophyllite-(Na).

Gyrolite

Chemical composition: $NaCa_{16}Si_{23}AlO_{60}(OH)_8$
$.14H_2O$ **Crystal system:** Triclinic **Mineral
habit:** Massive, concretionary, radiating
Cleavage: Perfect **Fracture:** Uneven
Hardness: 2½ **Specific gravity:** 2.45 to 2.51
Colour: Colourless, white **Streak:** White
Lustre: Vitreous; transparent to translucent

Gyrolite is a secondary mineral that occurs in
cavities in basalts and rhyolites affected by
hydrothermal fluids, together with various
minerals including apophyllite, okenite and
zeolites. It can form spherules or clusters up to
30 cm (12 in) in diameter.

▼ Gyrolite from
Talisker, Isle of Skye,
Scotland, UK.

Pyrophyllite

Chemical composition: $Al_2Si_4O_{10}(OH)_2$ **Crystal system:** Triclinic **Mineral habit:** Crystals tabular, frequently curved and distorted, radiating; usually foliated, fibrous, granular, compact **Cleavage:** Perfect **Fracture:** Uneven **Hardness:** 1 to 2 **Specific gravity:** 2.65 to 2.90 **Colour:** White, grey, yellowish, pale blue, green, brownish-green **Streak:** White **Lustre:** Pearly to dull; transparent to translucent

Pyrophyllite forms in medium-grade metamorphic rocks such as schists. Associated minerals include lazulite, kyanite, sillimanite and andalusite. It is also found in hydrothermal veins with micas and quartz. Pyrophyllite has a greasy feel and is very similar to talc. It is virtually insoluble in acids and will not melt in a flame, flaking when heated.

▼ Pyrophyllite from Bardin, North Carolina, USA.

Astrophyllite

Chemical composition: $K_2NaFe^{2+}_7Ti_2Si_8O_{26}(OH)_4F$
Crystal system: Triclinic **Mineral habit:**
Crystals bladed, frequently in groups with a
stellate (star-like) structure **Cleavage:** Perfect
Fracture: Uneven **Hardness:** 3 **Specific
gravity:** 3.20 to 3.40 **Colour:** Brown, bronze-
yellow to golden-yellow **Streak:** Yellowish
Lustre: Submetallic to pearly; translucent

Astrophyllite often occurs in hydrothermal veins
and in cavities in basalts and tuffs. It also forms
in other igneous rocks including syenites and is
associated with quartz, feldspars, micas, titanite,
zircon and riebeckite. Astrophyllite can be
dissolved with difficulty in acids, and melts in a
flame, producing a magnetic glass.

▼ Astrophyllite
from Khibiny Massif,
Murmansk Oblast,
Russia showing the
stellate radiating
crystal form.

Microcline

Chemical composition: $KAlSi_3O_8$ **Crystal
system:** Triclinic **Mineral habit:** Crystals
prismatic, large, tabular, very commonly
twinned; also massive, granular, compact
Cleavage: Perfect **Fracture:** Uneven
Hardness: 6 to 6½ **Specific gravity:** 2.54 to 2.57
Colour: White, grey, green, reddish, pink,
yellow **Streak:** White **Lustre:** Vitreous to
pearly; transparent to translucent

Microcline is a potassic feldspar and is dimorphous
with orthoclase, having the same chemistry, but
different internal structure. The striking green-
coloured variety is called amazonite. Microcline
is a very common constituent of many igneous
rocks, especially acidic granites and pegmatites, and
intermediate syenites. It also occurs in metamorphic
rocks, including schists and gneisses. Microcline
can form as large crystals a number of metres in
size. This mineral cannot be melted in a flame and
is insoluble in acids, except hydrofluoric acid.

▶ Microcline var.
amazonite on
smoky quartz from
Smoky Hawk claim,
Crystal Peak area,
Colorado, USA.

Orthoclase

Chemical composition: $KAlSi_3O_8$ **Crystal system:** Monoclinic **Mineral habit:** Crystals prismatic, tabular, large, very commonly twinned; also massive, granular, lamellar **Cleavage:** Perfect **Fracture:** Uneven to conchoidal **Hardness:** 6 **Specific gravity:** 2.55 to 2.63 **Colour:** White, colourless, grey, yellowish, reddish, greenish **Streak:** White **Lustre:** Vitreous to pearly; transparent to translucent

◀ Orthoclase from Crested Butte, Colorado, USA.

Orthoclase is a potassic feldspar which commonly occurs in acid igneous rocks, including granites, pegmatites and rhyolites, and in intermediate rocks such as syenites and trachytes. It also forms in many metamorphic rocks, especially gneisses and schists, and in detrital sedimentary rocks, having been derived from various original source rocks. Orthoclase is readily altered to clay minerals by hydrothermal fluids and chemical weathering. It is soluble only in hydrofluoric acid and can be melted in a flame with great difficulty, colouring the flame violet because of its potassium content.

Albite

Chemical composition: $Na(AlSi_3O_8)$ **Crystal system:** Triclinic **Mineral habit:** Crystals tabular, large, very commonly twinned; usually massive, granular, lamellar **Cleavage:** Perfect **Fracture:** Uneven to conchoidal **Hardness:** 6 to 6½ **Specific gravity:** 2.60 to 2.65 **Colour:** White, colourless; sometimes greyish, bluish, greenish, reddish **Streak:** White **Lustre:** Vitreous to pearly; transparent to translucent

▼ Albite from Morro Velho Mine, Minas Gerais, Brazil.

Albite belongs to the plagioclase series of feldspars and is the sodium-rich end-member of this solid solution series. Plagioclase feldspars can generally be distinguished from orthoclase feldspars by their repeated twinning. They commonly occur in many igneous rocks, including granites, pegmatites and rhyolites, syenites and andesites, and in metamorphic rocks such as gneisses and schists. Albite is also found in some hydrothermal mineral veins and in detrital sedimentary rocks. This feldspar can be formed by the process of albitization, which occurs when feldspars are altered by hydrothermal processes and albite is created. Albite is only soluble in hydrofluoric acid. It can be melted in a flame with difficulty, colouring the flame yellow due to its sodium content.

Anorthite

Chemical composition: $Ca(Al_2Si_2O_8)$
Crystal system: Triclinic **Mineral habit:**
Crystals prismatic, commonly twinned; also
massive, lamellar, granular **Cleavage:** Perfect
Fracture: Uneven to conchoidal **Hardness:** 6
to 6½ **Specific gravity:** 2.74 to 2.76 **Colour:**
Colourless, white, grey, reddish **Streak:** White
Lustre: Vitreous; transparent to translucent

▼ Anorthite from
Scotland, UK.

Anorthite is the calcium-rich end-member of
the plagioclase feldspar solid solution series.
This form of plagioclase occurs in higher
temperature igneous rocks than sodium-rich
albite. It is commonly found in basic igneous
rocks, including gabbros, dolerites and basalts,
and in some metamorphic rocks and meteorites.
Anorthite is soluble in hydrochloric acid. Two
intermediate members of the plagioclase feldspar
series are labradorite and andesine. The former is
well-known for its brilliant play of colours, called
schillerization, which is seen on cleavage surfaces.

▼ Labradorite
from Ylämaa area,
Lappeenranta, South
Karelia, Finland

Sodalite

Chemical composition: $Na_4(Si_3Al_3)O_{12}Cl$
Crystal system: Cubic **Mineral habit:** Crystals usually dodecahedral, commonly twinned; also massive, granular, nodular **Cleavage:** Poor
Fracture: Uneven to conchoidal **Hardness:** 5½ to 6 **Specific gravity:** 2.27 to 2.33 **Colour:** Colourless, white, greenish, blue, yellowish, reddish **Streak:** White **Lustre:** Vitreous to greasy; transparent to translucent

▼ Sodalite from Swartbooisdrift, Kunene Region, Namibia.

Sodalite is a feldspathoid mineral and gives its name to the sodalite group, which includes lazurite and hauyne. It forms in intermediate igneous rocks, especially nepheline syenites, in ejected volcanic rock, and in altered limestones. It can also be found in some meteorites. Sodalite dissolves in hydrochloric and nitric acids, making a siliceous gel. It melts in a flame, producing a colourless glass, and the flame is coloured yellow due to the mineral's sodium content.

Lazurite

Chemical composition: $Na_3Ca(Si_3Al_3)O_{12}S$
Crystal system: Cubic **Mineral habit:** Crystals rare, dodecahedral; usually massive, compact **Cleavage:** Imperfect **Fracture:** Uneven **Hardness:** 5 to 5½ **Specific gravity:** 2.38 to 2.45 **Colour:** Deep blue, azure blue, greenish blue, violet blue **Streak:** Blue **Lustre:** Resinous; translucent

Lazurite, also called lapis lazuli, is a member of the sodalite group of minerals and forms in limestones that have been affected by contact metamorphism. It is often associated with calcite. Lazurite frequently contains small grains of pyrite. It is soluble in hydrochloric acid and when held in a flame, it melts, producing a white glass. Lazurite is a feldspathoid mineral.

▼ Lazurite from Sar-e-Sang mine, Badakhshan, Afghanistan.

Nepheline

Chemical composition: $Na_3K(Al_4Si_4O_{16})$
Crystal system: Hexagonal **Mineral habit:**
Crystals prismatic, with hexagonal cross-section
and often with rough faces; usually massive,
compact, granular **Cleavage:** Indistinct
Fracture: Subconchoidal **Hardness:** 5½ to 6
Specific gravity: 2.55 to 2.66 **Colour:** Colourless,
white, grey, green, yellowish, reddish **Streak:**
White **Lustre:** Vitreous to greasy; transparent
to opaque

▼ Nepheline.

Nepheline occurs in various igneous rocks,
especially those of intermediate composition,
including nepheline syenites and pegmatites. This
feldspathoid mineral dissolves in hydrochloric
acid, producing a silica-rich gel. Nepheline can
be melted only with difficulty, and because of its
sodium content, it colours the flame yellow.

Scapolite group

Chemical composition: $(Na,Ca)_4(Si,Al)_{12}O_{24}Cl$
Crystal system: Tetragonal **Mineral habit:**
Crystals prismatic; also massive, granular,
columnar **Cleavage:** Distinct **Fracture:**
Uneven to conchoidal **Hardness:** 5½ to 6
Specific gravity: 2.50 to 2.62 (marialite), 2.74
to 2.78 (meionite) **Colour:** Colourless, white,
grey, bluish, greenish, violet, pink, yellowish,
brown **Streak:** White **Lustre:** Vitreous to
pearly or resinous; transparent to translucent

Scapolite is the name of a group of minerals
containing the marialite – meionite series.
Marialite is the sodium-rich end member and
meionite is calcium rich. Scapolite minerals
occur in rocks that have been altered by regional
metamorphism, such as schists and gneisses,
and in some contact metamorphic rocks. These
minerals also form in pegmatites and granulites.
They dissolve in hydrochloric acid and melt in
a flame. Orange-yellow fluorescence can often
be observed under ultraviolet light. Scapolite
minerals are occasionally used as gemstones.

▲ Scapolite.

Analcime

Chemical composition: $Na(AlSi_2O_6) \cdot H_2O$
Crystal system: Triclinic **Mineral habit:** Crystals trapezohedral, cubic; also massive, granular, compact **Cleavage:** Indistinct **Fracture:** Subconchoidal **Hardness:** 5 to 5½ **Specific gravity:** 2.24 to 2.29 **Colour:** Colourless, white, grey, yellowish, greenish, pink **Streak:** White **Lustre:** Vitreous; transparent to translucent

Analcime is a member of the zeolite group of minerals. It forms in basaltic lavas, with other zeolites, and can result from the alteration of nepheline and sodalite. It also occurs in detrital sediments, including sandstones and siltstones. Analcime dissolves in concentrated acids and melts to form a transparent glass, colouring a flame yellow due to its sodium content.

▼ Analcime from Dean Quarry, St Keverne, Cornwall, England, UK.

Heulandite-(Na)

Chemical composition:
$(Na,Ca,K)_6(Si,Al)_{36}O_{72} \cdot 22H_2O$ **Crystal system:** Monoclinic **Mineral habit:** Crystals tabular, trapezohedral; also massive, granular **Cleavage:** Perfect **Fracture:** Uneven **Hardness:** 3 to 3½ **Specific gravity:** 2.20 **Colour:** Colourless, white, grey, yellow, red, pink, brownish **Streak:** White **Lustre:** Vitreous to pearly; transparent to translucent

Heulandite-(Na) is the most common of a small subgroup of heulandite minerals, all of which belong to the larger zeolite group. Most heulandites form in cavities in basalts and andesites with a number of other zeolites and can also occur in gneisses and sandstones. They dissolve readily in hydrochloric acid, creating a silica-rich gel. When heated, they melt and give off water. The other members of the heulandite subgroup are named after the dominant chemical element in their composition. They are heulandite-(Ca), heulandite-(K), heulandite-(Ba) and heulandite-(Sr).

▼ Heulandite from Kilpatrick, Dumbartonshire, Scotland, UK.

Natrolite

Chemical composition: $Na_2Al_2Si_3O_{10} \cdot 2H_2O$
Crystal system: Orthorhombic **Mineral habit:**
Crystals slender prismatic, commonly acicular,
striated; also massive, granular, compact
Cleavage: Perfect **Fracture:** Uneven
Hardness: 5 to 5½ **Specific gravity:** 2.20 to 2.26
Colour: Colourless, white, grey; rarely yellowish,
reddish **Streak:** White **Lustre:** Vitreous to
pearly; transparent to translucent

This zeolite mineral occurs in cavities in
basaltic lavas with other zeolites. Natrolite
can also form by the alteration of sodalite
and nepheline in syenites, and of plagioclase
feldspars in aplites and dolerites. It can be
found in hydrothermal veins in serpentinites
too. Natrolite dissolves in concentrated acids
and melts to a transparent glass, colouring the

▼ Natrolite from Puy
de Marmant, Puy-de-
Dome, France.

flame yellow due to its sodium content. When
placed under ultraviolet light, it may fluoresce
an orange colour.

Stilbite-(Ca)

Chemical composition: $NaCa_4(Si_{27}Al_9)$
$O_{72} \cdot 28H_2O$ **Crystal system:**
Monoclinic **Mineral habit:** Crystals usually
in sheaf-like aggregates, and as cruciform
(cross-shaped) penetration twins; also massive,
globular, bladed **Cleavage:** Perfect **Fracture:**
Uneven **Hardness:** 3½ to 4 **Specific gravity:**
2.19 **Colour:** White, grey, yellowish, pink, red,
orange, brown **Streak:** White **Lustre:** Vitreous
to pearly; transparent to translucent

Stilbite-(Ca) belongs to the zeolite group
of minerals. It occurs in cavities in basaltic
and andesitic lavas and can also form in
granitic pegmatites, in hydrothermal veins in
metamorphic rocks and as deposits around hot
springs. Stilbite-(Ca) melts easily to a pale glass
and dissolves in hydrochloric acid. It forms a
series with the closely related stilbite-(Na).

◄ Stilbite from
Berufjord, Iceland.

Glossary

Accessory mineral A mineral that usually occurs in such small amounts in a rock that it is not relevant to the classification of that rock.

Acicular habit Needle-shaped mineral habit.

Acid rocks Igneous rocks that contain over 65% total silica and usually more than 10% quartz.

Adamantine lustre A very bright mineral lustre, like that of a diamond.

Amphiboles A group of ferro-magnesian silicate minerals, which are important as rock-formers, often containing hydroxyl molecules.

Amygdale An infilled vesicle (cavity) in lava.

Anhedral crystals Poorly shaped crystals, usually in an igneous rock.

Arenaceous Containing sand, usually referring to sedimentary rocks.

Aureole The zone around an igneous intrusion in which contact metamorphism has occurred.

Basic rocks Igneous rocks with a total silica content between 45% and 55%, and usually less than 10% quartz.

Batholith A discordant igneous intrusion of great size and often irregular shape.

Bedding Layering in sedimentary rocks.

Bladed habit Blade-shaped mineral habit.

Botryoidal habit A mineral habit, shaped like a bunch of grapes.

Chatoyancy Reflected light in some minerals resembling that in a cat's eye.

Clay minerals A term used to describe a number of common groups of generally aluminium-rich silicate minerals that form in microscopic layers and often have extremely small crystals.

Clast A fragment in a sedimentary rock.

Cleavage The way many minerals break related to internal atomic structure. Some rocks, especially slate, are also said to cleave.

Compound A material formed by two or more elements combining chemically; many minerals are chemical compounds.

Conchoidal fracture Mineral fracture which produces a curved, shell-like break; certain rocks, such as flint, can also show this fracture.

Concordant Following existing rock structures, such as bedding planes.

Concretion Rounded, discrete rock mass, often occurring in fine-grained strata.

Country rock Rock which is intruded by magma in a certain area and often metamorphosed.

Cross-bedding Sedimentary structure where layers are at an angle to the top and bottom of a bed.

Cryptocrystalline With minute crystals, requiring microscopic examination.

Dendritic habit Mineral habit, branching like a tree.

Detrital rocks Sedimentary rocks composed of fragments and grains.

Diagenesis Various processes which create rock from soft, loose sediment.

Dimorphous Denoting two distinct minerals with the same chemistry.

Discordant Cutting across existing rock structures.

Dull lustre Mineral lustre with poor reflectiveness.

Dyke Minor igneous intrusion; dykes are sheets of igneous rock that cut across existing structures such as bedding planes.

Earthy lustre Mineral lustre without reflectiveness.

Element A substance which cannot be broken down by chemical means into simpler substances.

Equigranular texture A rock texture with constituent grains or crystals of the same size.

Erosion Rock disintegration involving movement, as in a river or glacier.

Essential minerals Minerals in a rock, usually igneous, that define its chemical classification.

Euhedral crystals Well shaped crystals.

Evaporites Rocks and minerals formed by precipitation from saline or mineral-rich water as it dries out.

Extrusive rocks Igneous rocks which form on the Earth's surface; lavas.

Fault A break in crustal rocks where displacement has occurred.

◀ Azurite.

Feldspathoids Silicate minerals rather like those of the feldspar group, but containing less silica.

Ferromagnesian minerals Dense, dark-coloured minerals containing iron and magnesium.

Foliation planes Internal rock surfaces created by structural changes, such as slaty cleavage and schistosity.

Fossil Evidence of past life, preserved in rocks.

Gangue minerals Minerals of no economic value, often occurring in ore veins.

Glass Igneous rock formed by very rapid cooling.

Graded bedding Rock texture where coarser grains usually occur at the base of a layer, becoming progressively finer towards the top.

Graphic texture An intergrowth of quartz and feldspar, that looks somewhat like writing, usually occurring in granite.

Groundmass The body or mass of a rock, its matrix, often of uniform grain size.

Hackly fracture Mineral fracture with a rough surface.

Hemimorphism The property of a mineral crystal having differently shaped ends.

Hopper crystal Crystal with hollowed-out, often stepped, faces.

Hydrothermal fluids Hot fluids, often high-temperature water, commonly associated with igneous rocks.

Igneous rocks Rocks created by the consolidation of magma or lava.

Intermediate rocks Igneous rocks with a chemical composition between that of the acid and basic rocks, containing between 65% and 55% silica.

Intrusion A mass of rock which has invaded pre-formed rocks. Although most commonly relating to igneous rocks, salt domes can also be intrusive.

Lava Igneous rock, originally molten, on the Earth's surface.

Lustre The reflection of light from a mineral's surface.

Magma Molten igneous rock which can consolidate below the surface or be erupted as lava.

Mammillated habit Rounded mineral habit on a larger scale than a botryoidal habit.

Massive habit Shapeless mineral habit.

Mature rocks Sedimentary rocks containing a relatively high percentage of quartz.

Metallic lustre Mineral lustre like that of a metal.

Metamorphism The alteration of previously formed rocks by heat and/or pressure, without significant melting.

Metasomatism The processes by which minerals in rocks are chemically changed.

Nuée ardente A violent volcanic eruption producing an incandescent cloud of gas and lava droplets.

Ooliths Small, rounded, usually calcite-rich grains.

Orogeny A protracted period of mountain formation, usually associated with tectonic plate movement.

Pegmatite A very coarse-grained igneous rock.

Pelitic sediment Fine-grained sediment, including mud, shale and clay.

Phenocryst A relatively large crystal set in the groundmass of an igneous rock.

Phosphorescence The continued emission of light after the light source has ceased.

Placer deposit Minerals which are preferentially deposited because of their high specific gravity and resistance to erosion.

Playa lake A temporary lake in an arid region.

Pleochroism The way some minerals appear different colours when seen from different directions.

Pluton An igneous intrusion of great size, formed at considerable depth.

Porphyritic texture A texture of igneous rocks, with large crystals set in a finer groundmass.

Porphyroblastic texture A metamorphic texture with larger crystals (porphyroblasts) set in the rock's groundmass.

Primary mineral A mineral formed at the same time as the rock within which it is found.

Pyroclast Volcanic fragment.

Pyroxenes Important rock-forming ferromagnesian minerals lacking hydroxyl molecules.

Radiometric dating A method by which the absolute age of minerals and rocks is obtained by studying their radioactivity.

Reniform A kidney-shaped mineral habit.

Resinous lustre Mineral lustre with reflection like that of resin.

Scalenohedral A closed crystal form having faces which are triangles with unequal sides and angles.

Schistosity A type of wavy foliation commonly found in medium- and coarse-grained metamorphic rocks.

Secondary mineral A mineral formed after the formation of the rock in which it is found.

Sill A relatively small-scale concordant igneous intrusion.

Strata Beds (layers) in a sedimentary rock.

Texture The shape, size and relationship between the particles in a rock.

Thrusting Faulting with a low-angled plane, where one mass of rock moves over another.

Trimorphous Describing three distinct minerals with the same chemistry.

Tuff A rock composed of consolidated volcanic ash.

Twinning The sharing of a crystal face by two or more crystals of a single mineral.

Ultrabasic rocks Igneous rocks which contain less than 45% silica and are composed mainly of ferro-magnesian minerals.

Vein A joint or fault infilled with minerals.

Vesicle A cavity in igneous rock, usually lava, formed by a gas bubble.

Vitreous lustre Highly reflective mineral lustre, like that of glass.

Weathering Rock disintegration caused by *in situ* processes, without movement.

Xenolith A fragment of country rock caught up by, and held in, magma or lava.

Zeolites Hydrated aluminium silicate minerals characterised by reversible dehydration, where water can be gained by or lost from their composition.

Further information

Publications

Allaby, M. (2013), *A Dictionary of Geology and Earth Sciences*. Oxford University Press, Oxford.

Deer, W.A., Howie, R.A. and Zussman, J. (1992), *An Introduction to Rock-Forming Minerals*. Longmans, London.

Duff, P. (1993), *Holmes' Principles of Physical Geology*. Chapman and Hall, London.

Hamilton, W.R., Wolley, A.R. and Bishop, A.C. (1983), *Minerals, Rocks and Fossils*. Country Life, London.

Jerram, D. (2011), *Introducing Volcanology: A Guide to Hot Rocks*. Dunedin, Edinburgh.

Keary, P. (2003), *Penguin Dictionary of Geology*. Penguin, London.

Park, G. (2010), *Introducing Geology*. Dunedin, Edinburgh.

Pellant, C. (1992), *Rocks and Minerals*. Dorling Kindersley, London.

Pellant, C and H. (2014), *Rocks and Minerals*. Bloomsbury, London.

Roberts, W.L., Campbell, T.J. and Rapp, G.R. (1990), *Encyclopedia of Minerals*. Van Nostrand Reinhold, New York.

Smith, D.G. (ed) (1982), *The Cambridge Encyclopedia of the Earth Sciences*. Cambridge University Press, Cambridge.

Stow, D. (2005), *Sedimentary Rocks in the Field*. Manson, London.

Websites

Correct at time of printing

British Geological Survey
www.bgs.ac.uk

Geological Society of London
www.geolsoc.org.uk

Geologists' Association
www.geologist.demon.co.uk

Natural History Museum, London
www.nhm.ac.uk

The Smithsonian Museum
www.si.edu

US Geological Survey
www.usgs.ac

A source of useful information
www.mindat.org

A good source of mineral specimens
www.richardtayler.co.uk

Index

Page numbers in **bold** refer to the main entry for a rock or mineral, which usually includes an illustration. Page numbers in *italic* refer to illustration captions. For commonly occurring rocks and minerals only main references and captions are indexed.

C

Picture credits